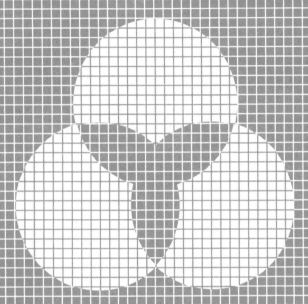

SCIENC

65 Experiments
That Introduce the Fun
and Wonder of Science

E WORKS

from the
Ontario Science Centre

Illustrated by
Tina Holdcroft

Addison-Wesley Publishing Company, Inc.

Reading, Massachusetts • Menlo Park, California • New York
Don Mills, Ontario • Wokingham, England • Amsterdam • Bonn
Sydney • Singapore • Tokyo • Madrid • San Juan

Library of Congress Cataloging-in-Publication Data
Main entry under title:

Scienceworks : 65 experiments that introduce the fun
  and wonder of science.

  "An Ontario Science Centre book of experiments."
  Summary: Provides instructions for experiments that
reveal a variety of scientific principles.
  1. Science—Experiments—Juvenile literature.
[1. Science—Experiments.   2. Experiments]
I. Holdcroft, Tina, ill.   II. Ontario Science Centre.
Q164.S299   1986   507'.8   85-26681
ISBN 0-201-16780-8

Cover design by Robin Mahler
Text design by Michael Solomon
        HIJ-AL-898
Eighth Printing, April 1988

Originally published in Canada in 1984 by Kids Can Press,
Toronto
First published in the U.S. in 1986 by Addison-Wesley
Publishing Company, Inc.

The Ontario Science Centre in Toronto, Canada, is a vast
science arcade. Its three connected buildings are filled with
more than 500 exhibits. You can play with them, test
yourself against them, try experiments with them. The aim
is to let you explore, experience, and enjoy science.

Although a visit to the Science Centre is unique, science
itself is all around us. And you don't need a laboratory full
of equipment to discover it. Now you can have the Ontario
Science Centre spirit at home with *Scienceworks*. All you
need for the experiments in this book is some curiosity and
ordinary household items.

These home-made science experiments have been
enthusiastically tried by thousands of readers. We hope you
have fun doing them, too.

CONTENTS

The Science Show

The Great Outdoors

Puzzlers and Mysteries

Energy Savers

Body Tricks

Things to Make

THE SCIENCE SHOW

No amount of science will enable you to pull a moose — or even a rabbit — out of a hat. But the next 14 experiments will help you amaze and surprise your friends.

TEST YOUR STRENGTH

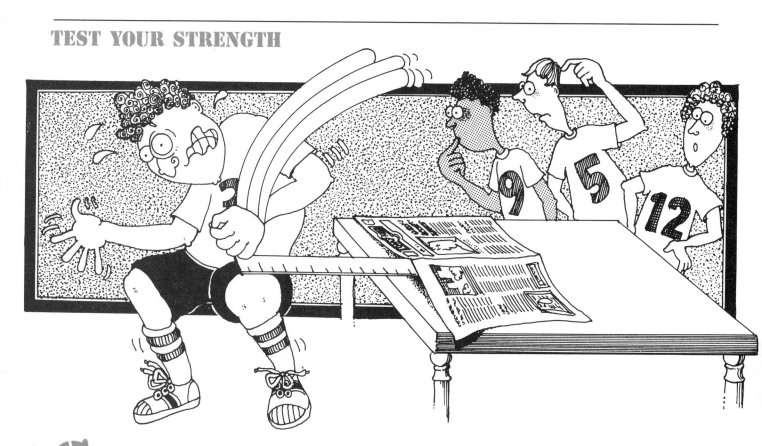

AN you hold up 45 kg (100 pounds) on one hand? It's easy. Just hold out your hand, palm up. There, you've done it. That's the weight of air. Here's a heavy air trick to play on your friends.

You'll need:
a sheet of newspaper
a strong ruler

1. Put the ruler on a table with one end over the edge. Cover the portion of the ruler on the table with the newspaper, as shown.
2. Tell your friend you're going to put a spell on the paper so you won't be able to lift it. As if performing magic, wave your hand over the paper, then bring your fist down sharply on the free end of the ruler.
3. Then have your friend try hitting the free end of the ruler. The paper still won't lift. In fact, if either of you hit the ruler too hard, it could break, and the paper wouldn't even be torn.

How does it work?
The weight of the air pressing down on the newspaper resists being squeezed up suddenly and holds the ruler to the table. The pressure of air is about 103 kPa (15 pounds on every square inch). It's hard to believe that air is that heavy!

INERTIA TRICK

HAVE you ever seen a magician whip a tablecloth out from under a full setting of dishes without even rattling a glass? There's no magic involved — only the clever use of science.

We don't recommend that you use dinner dishes for this. Instead, here's a safer — and cheaper — version of the tablecloth trick that will astound your family.

You'll need:
a heavy water glass
a strip of newspaper long enough to reach over the rim of the glass and hang over the edges
2 pennies
a ruler

1. Put one end of the strip of paper over the rim of the glass, as pictured.
2. Balance the two pennies on top of the paper, on the rim of the glass. Make sure the pennies are balanced on the rim and are not supported by the paper.

3. Lift the free end of the paper so that it is horizontal. Be careful not to move the coins.
4. Strike the paper with a ruler about 4 cm (1½ inches) from the edge of the glass. Speed is important. It might take a few tries before you get the hang of it, but before long, you should be able to whip the paper out from under the coins, leaving them balanced on the edge of the glass.

How does it work?
Inertia is a property of all things that makes them resist any change in their motion. When an object is standing still, it takes a force to get it moving. And once it is moving, it takes a force to stop it.

The heavier an object is, the more force — or time — it takes to change its momentum. That's the key to doing this trick. You can change the momentum of one part, the light paper, without changing the momentum of the other, the heavy pennies.

3

ICE FISHING

CAN you lift a floating ice cube out of a glass of water using just one end of a piece of string? Sounds impossible! It's not — if you know some basic chemistry.

You'll need:
a piece of string about 15 cm (6 inches) long
salt
a glass of cold water
an ice cube

1. Lay one end of the string across the top of the ice cube and sprinkle a little bit of salt on it.
2. Count slowly to 10, then gently lift the string. Presto! You've caught an ice cube.

How does it work?
Ordinary water freezes at 0° C (32° F). When you add salt to water, it lowers its freezing temperature — it has to get colder than 0° C to freeze. How much colder depends on how much salt is mixed in with the water. The salt you sprinkle on the ice cube lowers its freezing temperature and, since the ice cube can't get any colder than it already is, it starts to melt. A little pool of water forms on top of the ice cube and the string sinks into it. As the ice cube melts, it dilutes the salt/water mixture in the little pool; the freezing point starts to go back up again. The ice refreezes, trapping the string. As soon as the ice cube hardens, you can lift it by lifting the string. All this happens very quickly, of course.

An Icy Tip
Salt is useful for clearing ice from sidewalks and roads. But have you ever noticed that it doesn't work when temperatures get very, very cold? That's because if it's cold enough, the water will stay frozen. Even though the salt lowers the freezing point of water, the air temperature is so low that the ice doesn't melt.

MYSTERIOUS BREAKING STRING

 ERE'S another inertia trick that will puzzle your friends.

You'll need:

a medium-sized hardcover book

2 kinds of string, one stronger than the other, but both should break when you tug them. The weaker piece should be long enough to tie around the book at least three times.

1. Tie the strong string firmly around the middle of the book.
2. Cut the weaker string into three pieces, each long enough to tie around the book.
3. Tie one piece of weak string to the strong string on top of the book and another piece to the bottom, as shown.
4. Hold the book by the top string and pull down steadily and hard on the bottom string. (You might need to wear a glove to protect your top hand.) If you keep pulling steadily, the top string will break.
5. Replace the broken string with the third piece of weak string.
6. Hold the book in the air as before, and give a short sharp pull on the bottom string. Now it's the bottom string that breaks. Why?

How does it work?

When you pull slowly and steadily on the bottom string, you gradually pull the book down. This stretches the top string. Since the top string is bearing both the weight of the book and the strength of your pull, it eventually breaks.

When you pull suddenly and sharply, something very different happens. What makes the difference is the inertia of the book. Inertia is the tendency of all objects to stay at rest until some outside force makes them move. The bottom string breaks before the force of your pull can overcome the book's inertia. Since the book doesn't move, no extra force is placed on the top string, so it continues to hold.

If you hold the top string with your bare hand while you do these experiments, you'll feel the difference in the force that comes through to the top string — but it may pinch a bit!

5

 OOL your friends with this clever balloon trick. It looks like a snap — but it's not!

You'll need:
a balloon
a pop bottle

1. Push the deflated balloon into the bottle and stretch the open end of the balloon back over the bottle's mouth.
2. Challenge a friend to blow up the balloon. No matter how hard he huffs and puffs, he won't be able to do it.

How does it work?
As you inflate the balloon, it takes up more space in the bottle. But the bottle is already full — of air. Even though you can't see it, air takes up space. When you try to inflate the balloon, the air trapped inside the bottle prevents you from doing it.

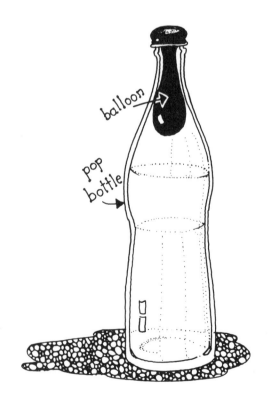

balloon

pop bottle

RED HOT TRICK

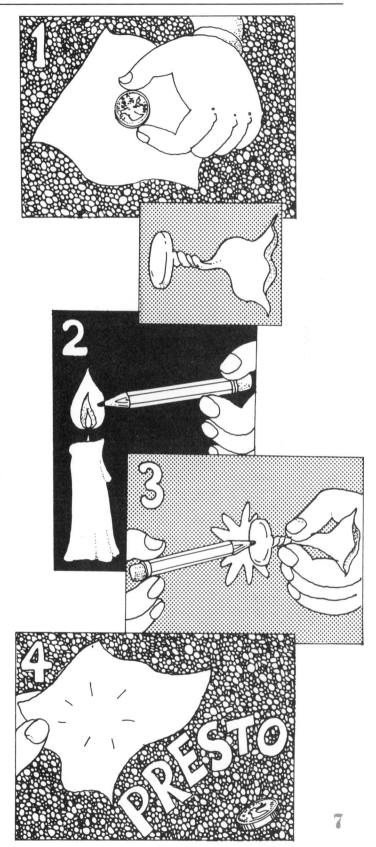

HAVE you ever tried to toast a marshmallow on the end of a wire hanger, instead of on the end of a stick? Before you've toasted many marshmallows, the wire will get too hot to hold. Here's a way to use your marshmallow-toasting experience to do an amazing feat. You can hold a piece of burning wood to a cloth without scorching it!

You'll need:
a quarter
an old cotton handkerchief or other piece of cotton
a pencil you no longer want
a candle in a candleholder

1. Place the quarter in the centre of the piece of cloth and twist the cloth so that it is stretched tightly over the coin, as shown. If the cloth is not pressed tightly enough against the coin, the material will scorch.
2. Light the candle. Hold the lead end of the pencil in the flame until the wood glows red hot.
3. Press the hot end of the pencil hard against the cloth where it covers the coin. Count to 10.
4. Remove the pencil, shake out the handkerchief, and blow away any loose ash. Presto! The cloth has not been burned.

How does it work?
Heat travels through different materials in different ways. Wood is a poor conductor of heat — heat does not travel quickly along a wooden stick. But metal is a very good conductor of heat — the heat travels quickly.

The metal in the coin is an excellent conductor of heat. It carries the heat from the smouldering wood right through the cloth so quickly that the heat has no time to scorch the material.

If the material did get scorched, you did not have the cloth pulled tightly enough against the coin. Therefore the coin was not able to conduct the heat directly from the smouldering wood.

7

TRY balancing a ruler on one finger. To do it, you have to rest the midpoint of the ruler on your finger. Here's a ruler balancing trick that seems to defy the laws of gravity.

You'll need:
a piece of string
a ruler
a hammer, preferably rubber-handled

1. Tie the string in a loop and slip it over the ruler and the handle of the hammer.
2. As shown in the illustration, position the hammer and ruler with the end of the hammer resting against the ruler. Your friends won't believe that the ruler can balance when only the top of it is on the table.

How does it work?

All objects have an imaginary spot called a centre of gravity that acts as if all the weight of the object were balanced there. The centre of gravity of a ruler is usually right in the middle of the ruler. But when you hang the heavy hammer from the ruler, you create a new system that has its centre of gravity near the head of the hammer so the ruler can balance from its tip.

Teeter-Totter Tip

If you've ever been on a teeter-totter with someone bigger than you, you know you're going to stay up in the air. The teeter-totter's pivot is right in the middle of the board you're sitting on, but there's more weight on one side of the board (your larger friend), than on the other (you). To get the teeter-totter balanced properly, you have to adjust the weights along the board so that the centre of gravity falls over the teeter-totter's pivot. The easiest way to do that is get your large friend to move towards you.

AN you pull two broomsticks together — even if your strongest friends are trying to hold them apart?

You'll need:
2 brooms
a strong rope or piece of clothes line at least 3 m (9 feet) long

1. Ask two strong friends to hold the broomsticks about 30 cm (1 foot) apart, and challenge a third friend to pull them together. When your friend has admitted defeat, you try it.
2. Tie one end of the rope securely around one broomstick.
3. Wind the rope around the two broomsticks, as shown.
4. Pull the free end of the rope. Presto! No matter how hard your two friends resist the pull, you can easily draw the broomsticks together.

How does it work?
By winding the rope around the broomsticks as you did, you used technology. Technology is the use of tools, materials, machines, and techniques to make work easier. In this case, wrapping the rope increases your pulling power. In fact, every turn of the rope approximately *doubles* your pulling power. But there's a hitch. Every time you double your pulling power, you must pull the free end of the rope twice as far.

This system of using a rope to increase your strength is a simple block and tackle. You can see the block and tackle being used on construction sites to lift heavy loads, in loading ships, and for lowering pianos from buildings. Remember, when Superman isn't available, technology is!

MAGIC WAND

JUST by combing your hair, you can make your comb appear to turn into a magic wand. Sound incredible? Try these experiments and see for yourself. It's best to try them on a dry day, and your hair should be clean.

You'll need:
a plastic or hard nylon comb
a piece of paper
a ping-pong ball

Magic Wand Trick #1
1. Turn on the cold water tap in the bathroom so that there's a thin stream of water coming from it.
2. Run your comb through your hair several times and quickly hold the comb close to the stream of water. Presto! Like magic, the stream of water bends towards the comb.

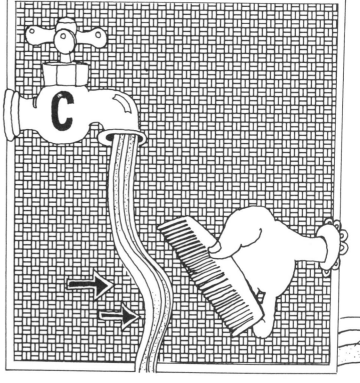

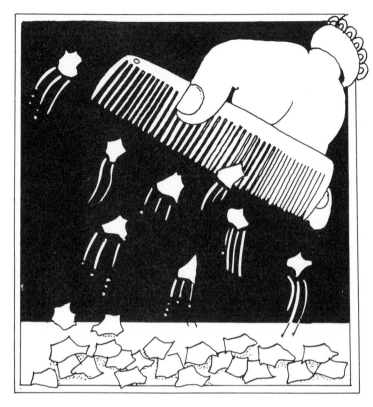

Magic Wand Trick #2

1. Tear up a piece of paper into tiny pieces.
2. Run your comb through your hair several times and quickly hold it over the paper. The bits of paper will jump up and stick to the comb.

Magic Wand Trick #3

1. "Charge" your comb by rubbing it on wool or some synthetic fabrics. (Test a variety of fabrics.) Rubbing it on a clean cat also works well.
2. Make a ping-pong ball follow you by holding your charged comb to it, and then slowly moving it forward. The ping-pong ball will roll along behind the comb.

How does it work?

All these tricks are the effects of static electricity, the same phenomenon that gives you a shock if you touch someone after shuffling across a carpet. When you run the comb through your hair (or shuffle your feet on the carpet), tiny particles called electrons move from one object to the other, leaving both with an electrical charge.

Electrically charged objects are able to attract things around them that have an opposite or neutral charge. They also repel things that have the same charge. That's why you may have noticed that, as you run the comb through your hair on a dry day, it leaves some hair standing away from your head — the hairs are charged and are repelling one another.

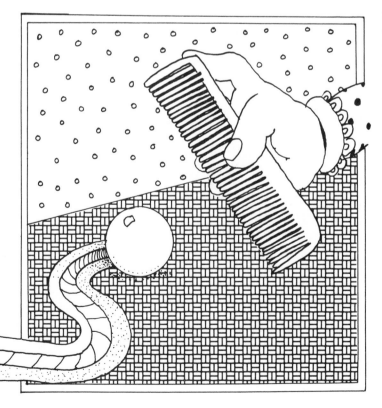

NOT-SO-THIN AIR TRICK

NEXT time you're looking for an empty glass, we guarantee you won't be able to find one. Why? All glasses, even the ones that look empty, are full of air. You can prove that air takes up space — and amaze your friends — with this easy trick.

You'll need:
a glass
a paper towel

1. Tell your friends that you are going to put the paper towel in the glass and then plunge the glass into the sink without getting so much as a drop of water on the paper towel.
2. Here's how to do it. Stuff the paper towel into the bottom of the glass. Make sure it's in there securely so it won't fall out when you turn the glass upside down.
3. Fill a sink with water. Hold the glass straight upside down and plunge it into the water.
4. Count slowly to 10, then carefully lift the glass out of the sink. Make sure you keep it perfectly straight at all times. Your friends will hardly believe their eyes when you pull a dry paper towel out of the glass.

How does it work?
Water could not get into the glass because it was full of air. And the air could not get out because it is lighter than water and couldn't escape under the rim of the glass.

paper towel

glass

EGG POWER TRICK

egg tape

EXT time someone's cooking with eggs around your house, save the eggshells so that you can astound your friends with this incredible stunt.

You'll need:
4 raw eggs
a small pair of scissors
masking tape
some books that are all about the same size

1. To crack the eggs and get four empty eggshells, gently break open the small end of each egg by tapping it on a table or counter.
2. Carefully peel away some of the eggshell.
3. Pour or scoop out the egg inside.
4. Put a piece of masking tape around the middle of each eggshell. This will prevent the eggshell from cracking when you cut it.
5. Carefully cut around the eggshell, through the masking tape, so that you have four half-eggshells with even bottoms.
6. Put the eggshells on a table, open end down, in a rectangle that's just a bit smaller than one of your books.
7. Lay a book on the eggshells. Do any of the shells crack?
8. Keep adding books until — CRRRACKK! How many books can you stack on the eggs? (For a real eye opener, weigh the books and see how many kilograms it took to break the shells.)

Why does it happen?
Each half of the eggshell is a miniature dome, and domes are one of the strongest shapes. Why? Weight on the top of the dome is carried down along the curved walls to the wide base. No single point on the dome supports the whole weight of the object on top of it. That's why domes are often used for big buildings that can't have pillar supports, such as hockey rinks and arenas.

Eggshell Trivia
Staff at the Ontario Science Centre in Toronto have shown that a single egg can support a 90 k (200 pound) person.

AMASING ROLLING CAN

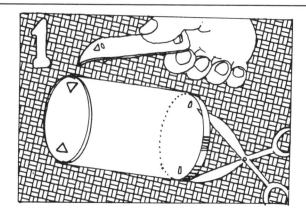

ERE'S a trick that uses rubber band power to make a can roll away from you — then back again all on its own.

You'll need:
a can "punch" opener
scissors
a coffee can or other can with a plastic lid
a long rubber band
a heavy nut or bolt

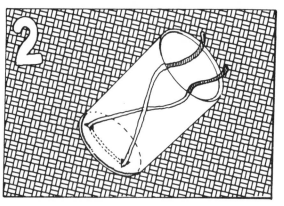

1. First, make your trick can. Use the can punch opener to make two holes on opposite sides in the end of the can. Punch matching holes in the plastic lid with a pair of scissors.
2. Cut the rubber band and feed it through the bottom holes, as shown.
3. Measure where the approximate centre of the band will be when it's stretched from end to end of the coffee can. Tie the nut or bolt to that spot so that when you're done, it will hang as shown in the illustration.
4. Thread the free ends of the band through the holes in the lid. Put the lid on the can and tie the rubber band ends together firmly on the outside of the lid.
5. Now your can is ready to perform. Roll the can away from you along level ground. When the can slows down, say, "Come to me." The can will stop — and then roll back towards you!

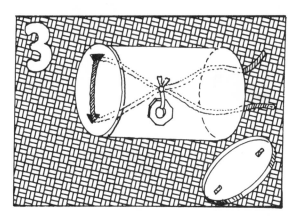

How does it work?
Rubber bands can store up energy and release it later. How? When you stretch or twist a rubber band, the band stores the energy it took you to stretch or twist it. When you let go, the energy is released. This trick makes use of the rubber band's ability to store and release energy. As you roll the can away from you across the ground, the weight inside the can causes the rubber band to twist, storing up energy. When the

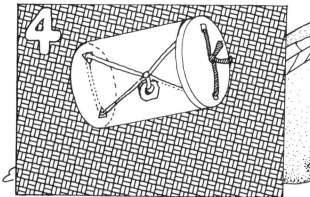

rubber band is tightly twisted, the can stops moving and starts to release its energy by rolling back towards you.

Rolling Up Hill
Make your returning can trick even more spectacular by rolling it downhill and having it apparently defy gravity by rolling back up to you. You could have it perform on a ramp by using a long board propped up at one end. Test your can first to see how far it will roll, and try to use a board long enough so the can will return before it reaches the end. Or, make a smooth junction with the floor by using cardboard so the can rolls off the ramp and comes back up it.

SCIENCE FRICTION

YOUR friends will think you're using a magic ruler for this trick. No matter how out of balance it looks, the ruler won't fall.

You'll need:
a metre (yard)-long ruler

1. Ask a friend to hold his hands about 60 cm (2 feet) apart with his palms turned inwards.
2. Place the ruler on top of his hands so that one end of the stick is very close to one hand and the other end is sticking out past the other hand.
3. Challenge him to move his hands together until the ruler becomes imbalanced and falls. No matter how many times he tries it, the ruler won't fall.
4. Ask him to bring his hands together at a point other than the centre. Can he do it?

How does it work?
Friction keeps the ruler from falling. What is friction? All objects resist moving across one another, and this is called friction. The heavier the weight of the object on top, the greater the friction between it and the object it's resting on. The stick is heavier at the long end so there is more friction between it and the hand on that side, making it harder for the hand under the long end to slide. While that hand is prevented from sliding quickly, the other hand slides to meet it. Like magic, the ruler stays balanced, and your hands always meet at the centre.

VERY TIGHT SQUEEZE

C HALLENGE your friends. Ask them to make a hole in a piece of notebook paper big enough to pass your whole body through without ripping the paper. Sound impossible? Here's how to do it.

You'll need:
a piece of paper (almost any size writing paper will do)
scissors

1. Fold the paper so that the two short ends meet.
2. Cut out a rectangle along the fold, as shown.
3. Make 13 cuts in the paper, as shown in the diagram.
4. Carefully stretch the paper out and you'll be able to climb through the hole. If the hole is a tight squeeze, try again on another piece of paper. This time, make more cuts. There must always be an odd number of cuts and they must follow the pattern.
5. Try it on a smaller piece of paper. How small can the paper be and still work?

How does it work?
What you've done is make it possible for the paper to stretch. If you carefully unfold the paper after you make the cuts and examine it, you'll see that there are points where the paper is strongly held together and other areas where it can pull away from the neighboring section. In a way, this is how the molecules of rubber alter their shape when you stretch them.

THE GREAT OUTDOORS

With sunshine and a lot of curiosity, the out-doors becomes a science laboratory for the next eight experiments.

EARTH SPEEDOMETER

WE all know that the Earth turns — that's what makes the sun appear to move across the sky. Here's a way to clock the Earth's speed using an easy-to-make solar speedometer.

You'll need:

a magnifying glass or any lens with one convex side, eyeglasses will work fine. (Find a lens that's gently rounded. Try looking through several, and pick one that puts the most distance between it and an object seen in focus through it.)

masking tape

a chair

a piece of white paper

a watch or clock with a second hand

1. Tape the handle of the magnifying glass to the seat of the chair so the lens extends horizontally over the edge, and place it in the sun.
2. Put the paper where the light passing through the lens shines on the ground. Raise the paper closer to the lens, or lift the chair to move the lens farther away from the paper, until you get a sharp circle of light, then use books or boxes to prop up the paper or the chair.
3. Draw a tight circle around the spot of light, then use your watch or clock to time how long it takes for the light to entirely leave the circle.

What's happening?

The spot of light is actually a tiny picture of the sun. When it moves fully out of the circle you have drawn around it, the Earth has travelled ½° of its 360° rotation. If you multiply the time it took for your "sunspot" to move that ½° by 720 and figure it out in hours, you'll find out approximately how long the day really is. Astronomers use atomic clocks to measure the exact length of the day.

Sun Catching
Solar energy is often collected with lenses or reflectors. You've seen how quickly the spot of sunlight moves, so you can imagine how difficult it is to keep the light falling directly onto the solar collectors. The most common solution to this problem is the use of motors to turn the solar collectors at the same speed as the Earth rotates, only in the opposite direction. This way, they point directly at the sun all day.

DOES it matter which way you plant a seed? After all, you want the plant's stem to grow up and the roots to grow down. Will the plant know which way to grow if you place the seed upside down? You can find out by making some glass gardens.

You'll need:
10 fresh beans (any kind, either from the supermarket or from a seed packet)
two wide-mouthed jars or glasses
a piece of blotting paper big enough to line the inside of both jars
paper towels

1. Soak the beans in water overnight.
2. Cut the blotting paper to fit snugly around the inside of each jar, as shown.
3. Stuff the middle of the jars with crumpled paper towels, then fill them with water and let the paper soak it up until it is saturated and will absorb no more. Pour off the remaining water.
4. Push five of the soaked seeds between the blotting paper and the glass in each jar, spacing them out evenly and keeping them near the top of the jars. Place the seeds in different positions — horizontal, vertical, and diagonal.

5. Put the jars where you can watch them for several days, but keep them out of direct sunlight. The blotting paper must be kept moist for the seeds to grow, so water the paper towels regularly. Over the next few days, you can watch the seeds germinate. Roots will grow from one end of each seed, and a stem from the other end, but no matter which way you placed the seeds, the roots will turn down and the stems will turn up. In less than a week, the seeds will have little green leaves.
6. After the seedlings have grown an inch above the top of the jars, lay one of the jars on its side. In a few days you'll see the stems are growing upward again, and the roots have bent to keep growing down!

How does it work?
There are growth hormones in plants that respond to the Earth's gravitational pull and make roots grow down and stems grow up. This response is called geotropism (Greek for "turning to earth"), and that's why you don't have to worry about planting seeds right side up.

SOLAR-POWERED clocks were invented thousands of years ago and you can make one. It's called a sundial.

The metric measurements cannot be converted into exact Imperial equivalents. For this project only, use the measurements for centimetres interchangeably with inches. A sundial built in inches will be more than twice as big as one built in centimetres.

You'll need:
2 pieces of heavy corrugated cardboard 20 cm (inches) square (Wood is better if you're going to leave it outside, but you'll need help sawing.)
a compass
a pen
a ruler
a protractor
scissors
white glue
masking tape

1. On one piece of cardboard 20 cm (inches) square, draw two diagonal lines from corner to corner, as shown. Where they meet is the centre of the cardboard.
2. Adjust the compass so that the point and the pencil are 9 cm (inches) apart, and draw a circle by putting the compass point in the centre of the cardboard. You should now have a circle 18 cm (inches) across.
3. Divide the circle in half and mark off 12 equally-spaced points around the circumference of one half. Number each point in this order, as shown: 6, 7, 8, 9, 10, 11, 12, 1, 2, 3, 4, 5. This is your dial.
4. Now you can start work on the triangular marker, called a gnomon (that's Greek for "one who knows"). On the second piece of cardboard, draw a line 8 cm (inches) long for the gnomon's base.
5. To make your sundial work properly, it must have the same angle as the latitude where you live. Use the Latitude Finder on the next page to get the correct angle for your gnomon. When you have it, use a protractor to mark off the angle at one end of the baseline, as shown.
6. Draw a line 20 cm (inches) long, from the end of the base through the mark, and connect the end of that line to the other end of the base. Cut out the gnomon.
7. On the dial, draw a line from the number 12 to the centre of the circle and mark a point 1 cm (inch) from the centre. Place the base of the gnomon along this line, with the measured angle touching the mark. Glue and tape it in place, and your sundial is ready.
8. Pick a spot for it where the sun shines all day, but — oddly enough — you'll have to wait until night to set it up. That's because the tip of the gnomon must point to the north, and the best way to do that is to line it up with the North Star (that's the bright one at the end of the Little Dipper's handle). Sight along the slant of the gnomon until the tip points to the star, then fix your sundial in place so it can't easily be moved.
9. To tell the time with your sundial, look at the shadow cast by the gnomon. The number on the dial where the edge of the shadow falls, is the correct time.

Why does the gnomon have to be at an angle?
Because the Earth is tilted on its axis, the sun appears to be lower on the horizon in the winter and higher in the summer. This apparent shift forms an angle with the horizon at sunset and sunrise. Wherever you are, that angle is equal to your latitude. If you build the gnomon of your sundial so that it forms the same angle with the base, the shadow will always fall in the same spot at the same time, all year round.

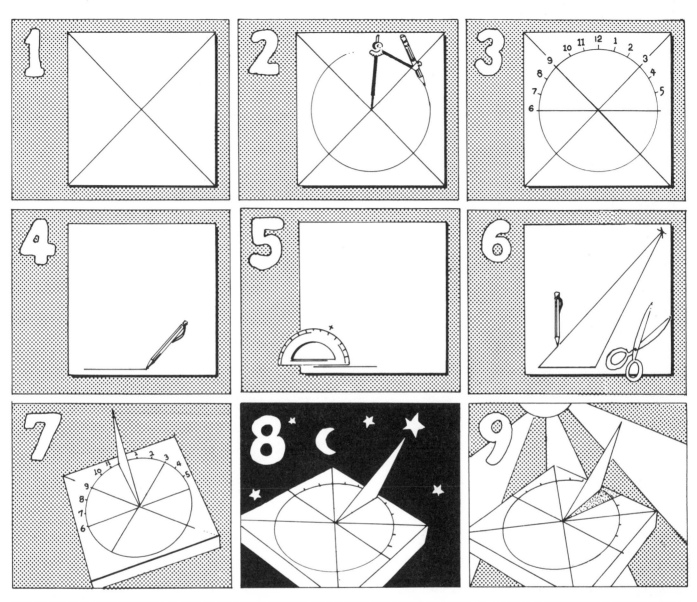

LATITUDE FINDER

City	Latitude*	City	Latitude*	City	Latitude*
Atlanta, GA	34 N	Detroit, MI	42 N	New York, NY	40 N
Baltimore, MD	39 N	Houston, TX	30 N	Philadelphia, PA	40 N
Boston, MA	42 N	Kansas City, MO	39 N	Phoenix, AZ	33 N
Chicago, IL	42 N	Los Angeles, CA	34 N	Saint Louis, MO	39 N
Cleveland, OH	41 N	Miami, FL	26 N	Seattle, WA	48 N
Dallas, TX	33 N	Minneapolis, MN	45 N	Washington, DC	39 N

NOTE: If you can't find your home town here, look in an atlas.   *rounded off to nearest degree.

WEATHER WEAR

WHAT does science have to do with fashion? Look around. As winter approaches and the weather gets cooler, people change into warmer, heavier clothes. These clothes have something else in common besides weight and warmth: they're usually much darker in colour than summer clothes. Navy blue, brown, dark greens, and dark reds are the colours of winter clothes, instead of pastels and white of summer. That makes good science sense, as this experiment shows.

You'll need:
2 white paper cups
a weather or cooking thermometer
some black paint

1. Paint the outside of one cup black.
2. Pour equal amounts of water (at the same temperature) into both cups and stand them side by side in the sun for about half an hour.
3. Test the temperature of the water in both cups. Which is warmer?

How does it work?

Don't be surprised if the water in the black cup is warmer. Dark colours absorb light and change it to heat. Light colours, on the other hand, act like reflectors and bounce light off. That's why it makes good science sense to wear dark, heat-absorbing colours in winter and light heat-reflecting colours in summer.

SOLAR WATER CLEANER

How can you get rid of the dirt in muddy water? Put it in the laundry? Strain it? The easiest way is to make this simple water purifier and let the sun do it.

You'll need:
a large pan or tub
a glass shorter than the pan
2 small, clean rocks
a piece of clear plastic food wrap big enough to fit over the pan
masking tape

1. Fill the pan to a depth of 5 cm (2 inches) with muddy water.
2. Set it where the sun will shine on it all day.
3. Put the glass right-side up in the centre of the pan and anchor it down, if necessary, by putting a small, clean rock in it.
4. Cover the pan with clear plastic. Pull it tight and tape it firmly to the pan.

5. Put a rock on the plastic wrap over the centre of the glass (don't let the rock touch the glass), then watch what happens. During the day, drops of clean water will form inside the plastic film and drip into the glass.

How does it work?
The sun's warmth heats the water, making it evaporate (turn into water vapour). When the vapour touches the cooler plastic wrap, it condenses back into water droplets. You have purified the water through a process called distillation. But what happened to the mud?

The dirt and stuff that make up mud don't evaporate at the same temperature that water does. So when the water vaporizes, it leaves the particles of mud behind. The water you collect in the cup has very few impurities.

Distillation is often used when the substances in a mixture have to be separated. For instance, it's one way of making fresh water out of salt water.

I F you left a pair of socks out in the yard this summer and forgot about them, what would you expect to find next spring when you went looking for them? That would depend on what they were made of — and on whether your dog found them first! To get an idea of what you might find, you can plant a reverse garden — reverse, because most people plant gardens to see things grow; you're planting this one to see things fall apart.

You'll need:
an old nylon stocking
some cotton cloth (an old sock or piece of towel will
   do, but make sure it's 100% cotton)
a piece of paper
some plastic wrap
some wool
a styrofoam or plastic cup
a piece of aluminum foil
an apple core

1. Dig a 12 cm (5 inches) deep hole for each item you're planting.
2. Pour enough water into each hole to thoroughly dampen the earth, then place one article in each hole and cover it with dirt. Be sure to put a marker over each item so you can find it again.
3. Leave the articles in your garden for 30 days and water them every day. At the end of that time, dig them up. How have things changed?

What happened?
Some of the things you "planted" have started to disintegrate. They are biodegradable — natural organisms can break them apart. What about the things that haven't disintegrated? Do you notice any similarities among them?

Be a Super Camper
Next time you're out camping, think of your reverse garden before you throw anything out. The two lists below show you which things are biodegradable and which aren't. Before you leave your campsite, you should bury the biodegradable things in a pit to speed up disintegration. Take the non-biodegradable garbage home with you.

Bury	*Take home*
food	plastic wrap
paper (you can	styrofoam cups
burn this)	or trays
	aluminum foil
	plastic bottles
	cans

COOL IT!

ow can you keep cool on those hot summer days? It's a snap with some simple science.

You'll need:
a pair of socks
some water
a warm, dry day

1. Soak one sock in water and then wring out the excess water.
2. Take the socks and your feet outside, put on the socks (one wet, one dry), and sit with your feet in the sun. Do you notice any temperature difference between your feet?

How does it work?
Your wet foot feels cooler thanks to evaporation. Evaporation is the process that uses energy to turn a liquid into a gas. Whenever liquid evaporates from a surface, heat is used up so the surface becomes cooler. And that's what happens to your foot. The evaporating water uses heat from the sun and your body. The result: cooler tootsies.

Evaporation is also what makes you feel cooler when a fan blows at you. The moving air makes perspiration evaporate more quickly. In fact, the cooling effect of evaporation is one of the reasons we perspire.

SOLAR COOKER

"IT'S hot enough to fry an egg on the sidewalk!" You've probably heard that expression, but have you ever tried it? If you have, you were probably disappointed. But don't give up. Here's a way to cook eggs and other foods in minutes, using just sunshine. You might need help making this solar cooker — all the measurements have to be exact — but it's worth the work.

This experiment can be done in any unit of measurement — centimetres, metres, inches, feet. In other words, one unit can be one inch or one centimetre or one-half inch, depending on the finished size you want. You can also make it bigger or smaller than the specified measurement. Just remember to keep the proportions the same as the ones described here.

You'll need:

a piece of corrugated cardboard 50 units square for the base

a pencil

a protractor

several smaller pieces of corrugated cardboard for the ribs

scissors

pins

glue

aluminum foil

food to cook

1. From corrugated cardboard, draw and cut out a circle with a radius of 50 units.
2. Inside the circle, draw and score (by cutting half-way through) a 45 unit radius circle.

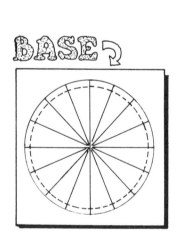

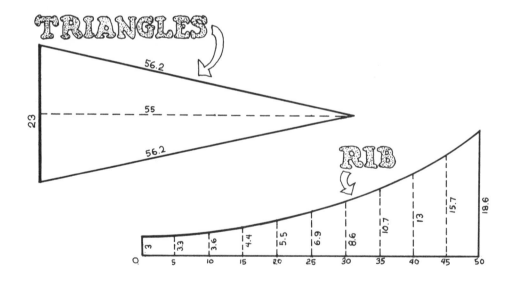

28

3. Using a protractor, divide the circle into 16, 22.5° sections. Then cut a slit from the outside to the scored circle.
4. Turn the circle upside down. Bend the scored ends upwards.
5. Cut out a rib on the other cardboard, using the measurements on the drawing. Trace it and cut out 15 more. You now have 16 identical ribs.
6. Glue the bottom edge and slip each rib into a slit. Hold them in position with pins until they're dry.
7. Cut 16 triangles of the size shown.
8. Cover each triangle with aluminum foil, shiny side up, and glue the foil down.
9. Glue eight triangles to the ribs. Each triangle fits over two ribs and covers one section. Wait for the glue to dry.
10. Cover the empty sections by gluing the eight remaining triangles to the ones already in place.
11. Hold the food above the centre of the dish. Move it up and down until you find the hottest point. A hotdog on a stick, or an egg in an aluminum pie plate, will cook in minutes.

How does it work?
The cooker you have made is a parabolic reflector. It collects the rays of the sun and focusses them at a single point above the centre of the dish. This cooker's focal point — the hot spot where the sun's rays are directly focussed and where food should be held — is about 40 units above the centre.

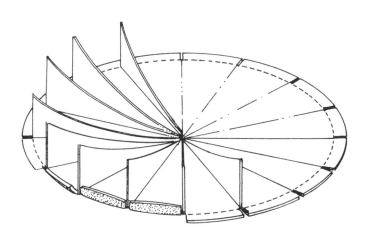

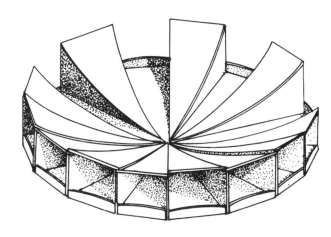

PUZZLERS AND MYSTERIES

Where does the wind come from? How can huge icebergs float? If you've ever wondered how and why things happen, read on. The next 17 experiments will help you answer some of your most-asked questions.

HOMEMADE RAIN

"IT's raining, it's pouring . . ." goes the old song. But where does all that rain come from? One way to find out is by making some rain in your own kitchen.

You'll need:
a large metal spoon or soup ladle
a kettle one-quarter filled with water

1. Put the spoon or ladle into the freezer to cool it.
2. When the spoon is ice cold, turn on the kettle. (Don't take the spoon out of the freezer until the water boils.) As the water in the kettle heats up, it turns into steam. Most people think the white vapour coming from the kettle is steam, but it's not. Real steam is invisible. If you look carefully — but not too closely — at the spout, you'll see a space between the kettle spout and where the white vapour starts. In that space is steam. As steam meets the air outside the kettle, it cools and becomes water vapour which is visible as a white cloud.
3. When the water is boiling, hold the cold spoon in the white vapour coming from the kettle's spout. Presto! In a few seconds it'll be "raining" in your kitchen.

How does it work?
Your cold spoon suddenly cools the water vapour that's coming from the kettle spout, making it condense into water and fall to the floor as "rain."

The Real Thing
Real rain is made in much the same way as homemade rain, but more gradually. Instead of a stove, there is the sun, which warms water in Earth's rivers, lakes, oceans, and even puddles. Fortunately for fish, frogs, and swimmers, not enough of the sun's heat reaches the Earth to make the water boil, but it is warm enough to allow tiny molecules of water to escape and rise into the sky. This is called "evaporation."

As the water-bearing warm air rises, it cools, and a cloud of water vapour forms, just like the cloud of water vapour formed when you boiled the kettle of water.

Cold air can't hold as much water as warm air. So when the air gets too cool to hold all the water vapour in it, some of the water falls back to earth as rain or snow. Then the cycle begins all over again.

F you put a tub of water on a scale and sat down in it, your weight would be added to the weight of the tub of water. But what if you just stuck your foot in the water without touching the bottom or sides of the tub? Would the scale register any added weight? Here's an experiment to help you solve this mystery.

You'll need:
2 water glasses, plastic are best
a strong ruler
a round pencil or a round dowel, the length of a pencil
tape

1. Tape the pencil or dowel to a table so it won't roll.
2. Balance the ruler across the pencil.
3. Put a glass on each end of the ruler. Pour water into each glass until they're about three-quarters full. Then adjust the amount of water in each glass until they balance evenly.
4. Stick your finger into the water in one of the glasses. Be sure you don't touch the bottom or sides

of the glass. What happens to the balance? Try putting your finger into the other glass and see if the same thing happens.

How does it work?
How can your finger make the glass of water heavier when you're not touching the glass? Try it again, and this time watch what happens to the level of the water in the glass when you put your finger in.

Your finger has taken the place of some of the water, and the displaced water piles up above its former level. If the displaced water just disappeared, the weight of the glass — with your finger in it — wouldn't change. Your finger not only takes the space of some of the water, it also "fills in" for the weight of that water. But, since the displaced water is still there, the glass weighs more, by the same amount as the weight of that displaced water.

Would it make a difference if you held a finger-sized piece of metal in the water? How about a finger-sized piece of wood?

33

MOEBIUS STRIP

YOU'VE probably heard the expression, "There are two sides to everything." But are there? You can find out.

You'll need:
several strips of paper about 25 cm (10 inches) long and about 2 cm (1 inch) wide
tape
scissors
2 different coloured crayons or felt markers

1. Make a circle out of one of the strips of paper and tape the ends together. Cut it in half around the middle. What do you get? Surprised? No, of course not.
2. Take another strip of paper. But this time, make a half-twist in it before you tape the ends together. Now cut this one in half down the middle. Oops! What happened?
3. Find out by making two more circles of paper, one with a half-twist and one without.
4. Test them with the crayons. Start with the plain strip. Put one crayon on the outside and trace it around until you come back to the beginning. Now put your other crayon on the unmarked side and draw a line until you come back to the beginning. Two sides, two colours, right? Now try the same thing with your twisted strip.
5. Try another test. Colour one edge of your plain circle. Try colouring one edge of the twisted one.

What's going on?
The circle with the half twist is the amazing one-sided, one-edged Moebius strip, named after Ferdinand Moebius, the mathematician who discovered it. It behaves in some surprising, but consistent, ways. When you cut your two-sided, two-edged strip in half, you get two pieces, with a total of four edges and four sides. Your cut-in-half Moebius strip is still in one piece, but how many edges and sides does it have?

If you cut an ordinary strip of paper around the middle twice, you'll end up with three separate pieces, with a total of six edges and six sides. Try cutting your Moebius strip around the middle twice. Surprised? See if you can guess how many sides and edges you have before checking them out.

WATER FLOWING UP-TREE MYSTERY

N EXT time you're flopped out under a shady tree, ask yourself this: how does water flow up from the tree's roots to its leafy crown? Here's an experiment to help you solve the mystery.

You'll need:
a cup, half-filled with water
some blue or red food colouring
a stalk of celery with some leaves on it

1. Mix a teaspoon of the food colouring into the water.
2. Cut the celery stalk about 2 cm (about 1 inch) from the bottom to expose a fresh end and stand the stalk in the water.
3. Leave the celery in the water for an hour or two and you'll see the dye gradually colouring the leaves.
4. When the colour has spread to the tips of the leaves, take the celery out of the water and cut across the stalk. You'll see a row of tiny circles outlined in colour — they're the cut ends of fine long tubes that travel the length of the stalk. The coloured water travelled up those tubes. Trees have similar tubes running up their trunks.

How does it work?

What makes the water climb the trees is still somewhat of a mystery. But scientists think it all depends on the special properties of water, and on the fact that the tubes are porous and very narrow. As the tubes spread out into the leaves, heat from the sun evaporates the water molecules at the top. Because water tends to climb a short way up the walls of certain substances (like drinking glasses, for instance), the next molecules in line move up after those that evaporate. Water molecules always hold tightly together, and when they're squished into very narrow tubes, they grip even more tightly, with enough strength to pull all the following water molecules along behind them. So as the molecules at the top move up, the whole chain moves up the tree. This only works, however, if the tubes are full of liquid to begin with, so trees and other plants have liquid-filled tubes from their earliest days as seedlings.

TOO MUCH PRESSURE

ALTHOUGH we can't see it, we live at the bottom of an ocean of air 483 km (about 300 miles) high! All that air presses down on us and everything around us. The reason we're not crushed from the great pressure of all that air is that there's also air inside us and under us that's pressing outwards and upwards with the same pressure. So it equals out. Here are two experiments that let you see that air pressure exists.

Air Presses Up

You'll need:
an ordinary water glass
a piece of stiff, flat cardboard

1. Fill the glass with water right to the top.
2. Slide a piece of cardboard over the top of the glass, making sure no bubbles of air are left in the glass.
3. Hold the cardboard tight against the glass and turn the glass upside down over the sink. Take your hand away from the cardboard. What happens?

How does it work?

What keeps the cardboard in place is the pressure of the air pushing up from the outside. This air pressure is greater than the weight of the water pushing down on the cardboard from inside. As long as the cardboard does not get soggy and sag, it will stay in place by itself. If the cardboard was not firm and flat to begin with, it will let air in and water out, so this experiment won't work.

Blowing Air Away

You'll need:
2 long strips of paper, any size

1. Holding the two strips of paper a few centimetres (about 2 inches) apart, dangle them in front of you.
2. What do you think will happen if you blow steadily between the strips? They'll be blown apart, right?

How does it work?

Why do the strips move *together* when you blow between them, instead of *apart*? You're blowing the air away from between them and lowering the air pressure. The pressure of the air outside the strips becomes greater than the air between the strips, so the pieces of paper are pushed together.

AIR

I F your fish tank sprung a leak, would the water spurt farther if the leak were nearer the top or the bottom of the tank? Here's how to find out.

You'll need:
an empty milk carton
a pen or pencil
a piece of adhesive tape the length of the carton

1. Use the pencil to punch three or four holes, one above the other, on one side of the carton. Make the top hole at least 3 cm (1 ¼ inches) from the top of the carton.
2. Cover the holes with the adhesive tape.
3. Fill the carton with water.
4. Put the carton in the sink or bathtub. Strip off the tape quickly. Which stream of water travels farthest?

How does it work?
Water pressure is the key to the answer. The water near the bottom of the carton has the force of all the water above it pushing it out. The water near the top has very little water — and therefore pressure — above it.

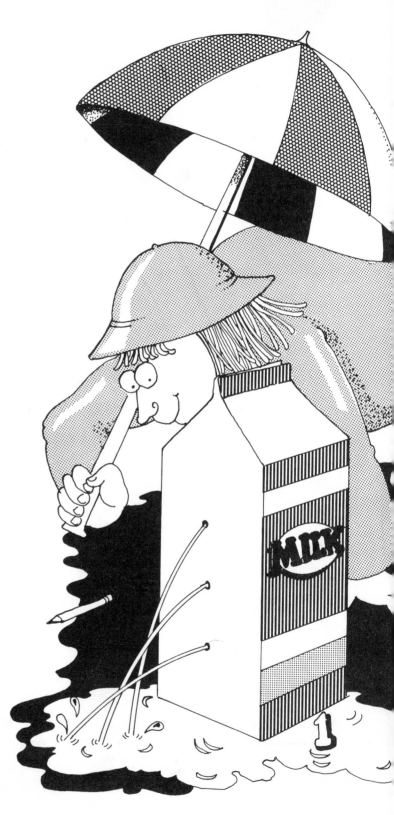

I F the *depth* of the water makes the pressure greater, what about the amount of water?

You'll need:
a small, frozen juice can
a large coffee tin or litre (quart) size tin
adhesive tape
an awl, or other instrument for making a hole in tin

1. Punch a hole 2-3 cm (about one inch) from the bottom of each can.
2. Cover the holes with a strip of tape.
3. Fill each can with water to the same depth (e.g. 5 cm [2 inches] deep in each can). The larger can will take much more water to fill the same depth as the small can.
4. Put both cans in the sink or bathtub. Pull off the tape at the same time. Which stream goes farther?

WHERE DOES THE WIND COME FROM?

ID you ever wonder what makes the wind? Here's an experiment that turns an ordinary light bulb into a wind-making machine.

You'll need:
a lamp
talcum powder

1. Remove the lamp's shade and turn on the lamp.
2. When the bulb is hot, sprinkle a tiny bit of talcum powder just above it, and watch what happens.

How does it work?
The powder is carried upwards by a rising current of warm air, or wind, warmed by the light bulb. Real wind starts when the sun heats the earth. As the earth gets warm, it heats the air just above it. This hot air expands, making it lighter. The warm, light air rises, leaving room for heavier, cooler air to move in and take its place. This movement of air is what we call wind.

YOU'LL need:
a light bulb
a pencil
a piece of paper
scissors

1. Cut a spiral out of the piece of paper, as shown.
2. Balance the centre of the spiral on the point of a pencil. You may have to make a small indentation in the paper to keep it from slipping off, but be careful not to make a hole.
3. Turn on the light bulb and wait a few minutes until it is hot. Then hold your pencil with the balanced spiral just above the bulb. What happens?

How does it work?
The spiral started to spin because the hot light bulb warmed the air around it. This hot light air started rising, creating a mini-wind that spins the spiral like the real wind spins a pinwheel.

cut paper spiral

2.

HOT SIPS

HY do the first few sips of cocoa always seem hotter than the later ones? Does your mouth get used to the heat? Does the drink cool off as you sip it? To solve the mystery, try this.

You'll need:
a piece of string about 30 cm (1 foot) long
a small bottle
a large jar
food colouring

1. Tie the string around the neck of the small bottle.
2. Fill the large jar with cold water.
3. Fill the small bottle with hot water and quickly stir in enough food colouring to make a strong colour.
4. Use the string to gently lower the small bottle into the large one filled with cold water. Don't let the small bottle tip. As the bottle drops, it will release a coloured fountain of hot water. Even after the bottle is settled on the bottom of the jar, coloured water will rise out of it. Soon all the coloured water will be floating at the top of the jar.

How does it work?
Water expands and rises when it's heated. This is what makes the hot, coloured water rise to the surface. It's also the reason why the top layer of cocoa is hotter than the rest.

Do you think this holds true for cold drinks, too? Can you think of a way to find out if cold drinks are colder on the bottom?

Here's a hint: try using an ice cube made of vegetable-dyed water.

WATER WONDER

WHY is a full glass of water like a bus at rush hour? Because you can usually put more into it! Surprise yourself by finding out how much more you can put into a full glass of water.

You'll need:
a glass
some food colouring
some coins

1. Fill a glass right to the top with water coloured with a few drops of food colouring.
2. Start gently dropping coins into the water. (It's best to hold the coins on edge and slip them into the water.) You'll notice that the top of the water bulges out above the top of the glass. How many coins can you add before the water overflows?

How does it work?
Water molecules have a strong attraction for one another. Inside the glass, the molecules that are surrounded by other molecules are attracted in all directions. But the molecules at the surface have no water above them, so they are strongly attracted downwards by the molecules below them. These attractive forces are strong enough to keep the water from spilling over the top of the glass, even when the level rises quite a bit beyond it. But eventually the volume of water above the rim of the glass becomes too great for the surface tension to hold, and the water will spill.

Minipuzzler
What happens when you add a few drops of dishwashing liquid to the water in the glass? Can you add more pennies than before, or fewer?

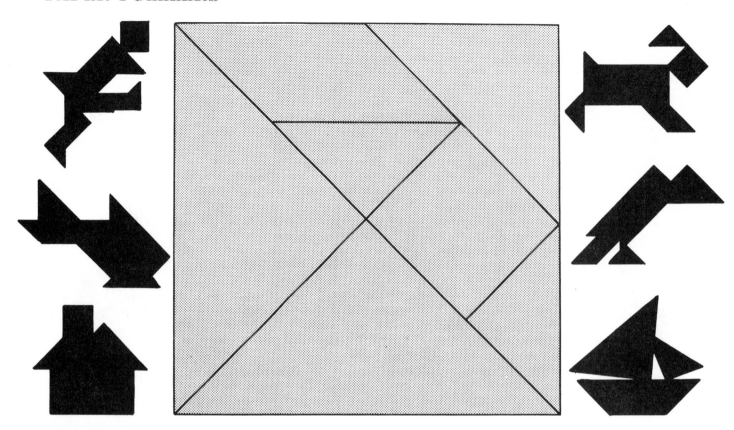

HESE two puzzles are guaranteed to give your brain a good workout. If you get stuck, try them out on your friends and family.

You'll need:
paper
a pencil
scissors

Rubic's Cube, Move Over
This puzzle is a match for the Rubic's Cube.
1. The 12 shapes shown on the next page are all the possible arrangements of five squares. Trace them onto a piece of paper and cut them out.
2. Try to fit them into a rectangle six squares high by 10 squares long, using all 12 shapes. Stumped? You'll find one solution on page 52. But according to a computer, there are 2338 other ways.
   If you figure out all 2338 other ways, you may be eligible for *The Guinness Book of Records*.

Tangram Puzzle
Here's how to make a challenging geometrical puzzle called a tangram.

1. Trace the square above onto a piece of paper.
2. Cut out along the black lines to make seven pieces, or "tans."
3. Choose one of the drawings and try to make it using all seven tans. Stumped? Keep trying different combinations before you check your solution on page 52. Once you've mastered all of the tangram puzzlers given, see how many figures you can create with the tans.

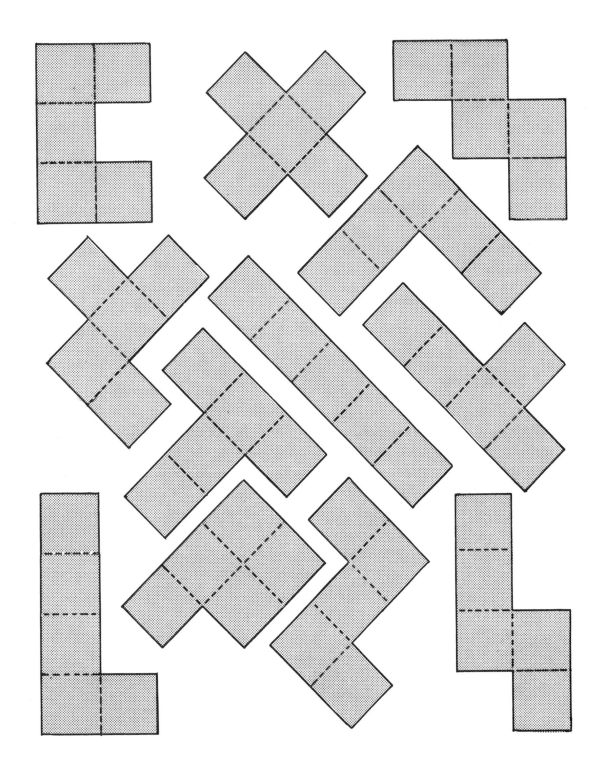

WHY can't you sink an ice cube? It's because an amazing thing happens when water freezes.

You'll need:
a glass jar
water
a freezer (the freezing compartment of a
   refrigerator is fine)

1. Place the empty glass jar in the freezer. Set it where it will stand upright and stay steady.
2. Fill the jar to the brim with water.
3. Close the freezer and come back in three hours and see what has happened.

What's going on?
Almost everything on earth contracts, or shrinks, when it freezes. Amazingly, water does just the opposite — it expands. When water freezes, its tiny molecules arrange themselves in an ordered array called a crystal, or lattice. This arrangement of the molecules takes up more space than the haphazard arrangement found in liquid water.

   This is the reason for another amazing fact about water — the same volume weighs less when frozen than when liquid. That's why ice cubes float in a glass of water and why icebergs float in the ocean.

How do movies move? Here's how to make a toy, called a thaumatrope, that will reveal the secret.

You'll need:
a pen and paper
glue
some stiff cardboard
scissors
string

1. Trace the circles illustrated and their drawings onto a piece of paper. Be sure to mark in the dots (•) and stars (*).
2. Cut out the two paper circles and one cardboard circle the same size.
3. Glue the paper circles on opposite sides of the cardboard disk. Make sure that stars are opposite each other.
4. Carefully punch tiny holes where the dots appear and thread string through them, as shown.
5. Twirl the disk by first twirling the string. As you twirl, watch the bird and cage. Like magic, the bird will appear to be inside the cage.

How does it work?
Your eye holds onto the image of the bird for a split second after it's gone. By that time the cage is in sight, and you seem to see the bird in the cage. The same thing happens at the movies. If you look at a piece of movie film, you'll see that it is a series of pictures separated by black spaces.

When a picture is flashed on the screen, your eye retains the image for a fraction of a second afterwards, so you don't see the moments of darkness between the frames. This is called persistence of vision. When a series of still pictures of moving objects is flashed before your eyes very quickly, you see them as uninterrupted movement. In order for this to work, at least 16 images must be flashed on the screen every second.

When you go to the movies, you see 24 images flashed on the screen every second, separated by intervals of darkness of the same length. So, in fact, you spend about half of a movie in the dark!

47

WHODUNIT?

Do you ever suspect that someone's been in your room without asking permission? Here's a simple fingerprinting experiment to help you finger the culprit.

You'll need:
a stamp pad
a piece of white paper
talcum powder
a fine paintbrush
clear cellophane tape
shiny black paper (dark blue will work too, but not
    as well)

1. Before anyone becomes suspicious, get a record of all possible suspects' fingerprints. Have each person press the pads of his or her fingers, one at a time, onto an ink pad, then roll the inky fingers on a piece of paper laid flat on a hard surface. In order to get the entire pattern, you have to roll the finger instead of just pressing it down flat.
2. Examine the prints. You may already know that each person's fingerprints are different from everyone else's, but does each person have the same print on all their fingers?
3. Now it's time to see if you can match a suspect's fingerprints with fingerprints in your room. Dust the talcum powder lightly on several hard surfaces in your room such as a desk, a light switch, and the doorknob.
4. Blow on the talcum powder gently so that most of the powder blows away, except for spots where the powder sticks to fingerprints and other greasy marks.
5. To reveal the prints, brush the powdered spots very lightly with your fine paintbrush until the pattern shows. It may take some practice until you learn to brush enough to reveal the pattern without damaging the print.

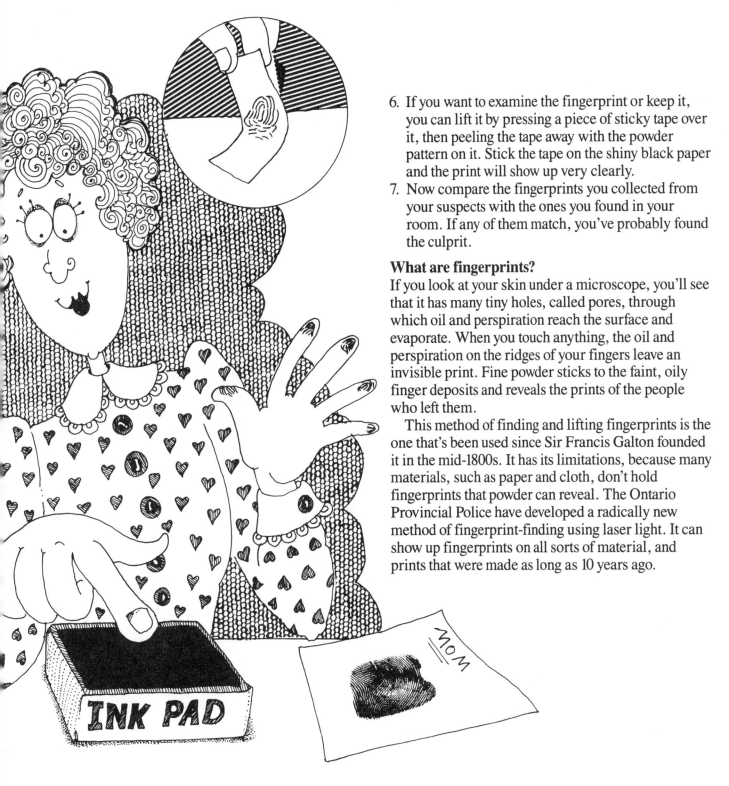

6. If you want to examine the fingerprint or keep it, you can lift it by pressing a piece of sticky tape over it, then peeling the tape away with the powder pattern on it. Stick the tape on the shiny black paper and the print will show up very clearly.

7. Now compare the fingerprints you collected from your suspects with the ones you found in your room. If any of them match, you've probably found the culprit.

What are fingerprints?

If you look at your skin under a microscope, you'll see that it has many tiny holes, called pores, through which oil and perspiration reach the surface and evaporate. When you touch anything, the oil and perspiration on the ridges of your fingers leave an invisible print. Fine powder sticks to the faint, oily finger deposits and reveals the prints of the people who left them.

This method of finding and lifting fingerprints is the one that's been used since Sir Francis Galton founded it in the mid-1800s. It has its limitations, because many materials, such as paper and cloth, don't hold fingerprints that powder can reveal. The Ontario Provincial Police have developed a radically new method of fingerprint-finding using laser light. It can show up fingerprints on all sorts of material, and prints that were made as long as 10 years ago.

TIED UP IN KNOTS

HAVE you ever tied your dog's lead to a pole, gone inside for a minute, and returned to find him gone? Or have you ever put up a tent only to have it collapse in the middle of the night?

If you've had these problems, you need a bit more knot know-how. Once you do, everything should go without a hitch . . . or maybe even with one.

All you need to know
Knots are classified into three general categories:
hitches (which tie a rope to an object)
bends (which tie two ropes together)
splices (which permanently join the ends of two
    ropes, or form a loop in a rope)

Here are a few hitches and bends you might find useful.

You'll need:
two pieces of strong rope
a piece of wood, or pole, or the back of a chair

Bowline hitch
If you want a really strong knot that never slips, this is one of the best. Use it to tie a lion to a lamppost.

Reef, or square, knot
This bend is probably the first knot you ever learned, but it's often misused. It's great for tying the two ends of a rope together — when you're tying parcels, for instance. But it will come undone very easily if it's used to tie two rope ends together when there will be different amounts of pull on each end, or if you're trying to tie cords of different sizes or materials.

Sheet bend
A generally useful knot for tying rope ends together, it is the best knot to use when joining ropes of different sizes.

Bowline Hitch

Reef or Square knot

Sheet bend

SOUNDS LIKE FUN

ID you ever wonder how sound travels? Here's a neat experiment that lets you see sounds on the move.

You'll need:
2 glasses about the same size and shape
a pencil
a piece of fine wire long enough to rest on the rim of one glass, as shown

1. Half fill the two glasses with water.
2. Tap the first glass with a pencil. You'll hear a musical note. Try to produce the same note on the second glass. You'll have to add or subtract water to get the second glass to make the same note.
3. Set the two glasses about 10 to 12 cm (4 to 5 inches) apart and rest the fine wire across the top of the one farthest from you.
4. Now tap the nearest glass and you will see the wire move slightly on the other one!

How does it work?
When you tap the first glass with a pencil, you start it vibrating. Although the vibrations are too small to be seen, they're strong enough to push the air forward in waves, like a stone thrown in water pushes out ripples. These sound waves cause similar vibrations in the other glass. It's these vibrations in the second glass that make the wire move. If you keep tapping the first glass, you can make the wire move to the edge of the second glass and it will eventually fall off.

Hear, hear!
Your ear picks up vibrations much like the wire does. Inside your ear is a sensitive piece of tissue called an eardrum. It vibrates when sound waves hit it. These vibrations are transmitted to your brain via your middle and inner ear. There the vibrations are "decoded" into sounds.

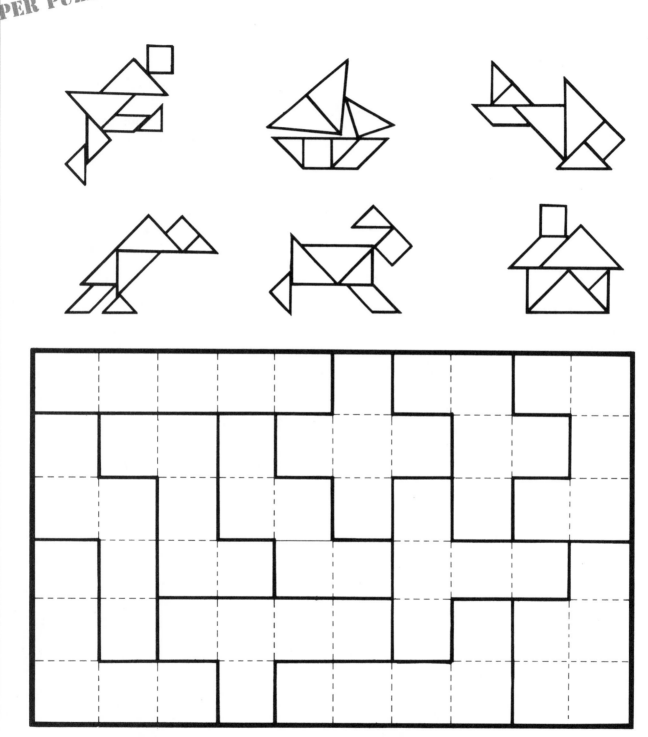

ENERGY SAVERS

Saving energy makes good sense and often saves cents. The next five experiments are guaranteed to leave you with extra energy, if not pennies in your pockets.

NEVER FAIL LID REMOVER

F you've ever tried to unscrew a tight metal cap on a jar, you know what happens. You twist and strain, and the jar gets passed from person to person until finally someone manages to undo it. If you want to save energy (yours), just hold the jar lid under hot water for half a minute. Then a little twist of the wrist and it'll come right off!

How does it work?

Most things expand as they get warmer. When you hold the jar lid under hot running water, the lid gets warm and expands faster than the glass jar underneath, making it turn more easily. Metal is a better conductor of heat than glass. Check it yourself by putting a drinking glass and a spoon in a pan of hot water, leaving the top of the glass and the spoon handle above the surface. After a minute, feel them to see which is warmer.

Not only does metal heat and therefore expand faster than glass, but it also expands more than glass.

RAMP MAGIC

EXT time you have to lift something heavy, try using an energy-saving ramp. Here's an experiment to show how a ramp can help you.

You'll need:
a thin rubber band
a rock about the size of your fist
a piece of string long enough to tie around the rock
a ruler
3 books

1. Wrap the string around the rock and tie the rubber band to it so that you can pull the rock.
2. Stack the books on top of one another and lean the ruler on top of them, as shown.
3. Use the rubber band to pull the rock up the ruler. Notice how far the rubber band stretches.
4. Remove the ruler and lift the rock up onto the books using the elastic. Does the rubber band stretch farther this time?

How does it work?
The longer the band gets, the more force you're using to move the rock from the ground to the top of the pile of books. It doesn't matter whether you use a ramp or lift the rock straight up, you're doing the same amount of work, but the length the rubber band stretches will show you the difference in the force needed to do that work. Now use a different length of ruler as a ramp and see what happens. When you use a ramp, its size depends on the size and weight of the object you are trying to move.

ENERGY-SAVING AQUARIUM DRAINER

To clean an aquarium, you first have to get rid of the water. And if your aquarium is a big one, that can be a tough job. One way to get the job done — and save energy — is to use a siphon. Here's an experiment to show you how it works.

You'll need:
2 glasses
a thick book
some rubber tubing
a table

1. Fill one of the glasses almost to the top with water and stand it on the book.
2. Put the empty glass on the table beside the book.
3. Fill the tubing with water and pinch the ends tightly so the water doesn't run out. (Use clothes pegs if you find it hard to hold both ends of the tube closed.)
4. Put one end of the tube underwater in the full glass, and let go of that end. Make sure the end doesn't flop out of the water.
5. Bend the tubing and put the other end in the empty glass.
6. Let go and watch the water flow.

7. Try raising the lower glass, without letting the tube come out of the water at either end. Does the water stop flowing at any point? What happens if you raise the lower glass above the one on the book?

How does it work?

Gravity pulls the water down on the lower side of the tube. At the same time, air pressure pushes the liquid through on the high side. Together, they create the flow of water through the siphon. When the water is at the same level in both containers, gravity is exerting the same pull on both sides of the tube and the flow stops.

Do you think the siphon would work if the tube weren't filled with water? Try it. With air in the tube, water is prevented from rising to the top of the high side. You can start a siphon working by putting one end in a container of water, then sucking on the other end until the water is almost all the way through the tube. Quickly take the end out of your mouth and hold it, pointing down, at any point lower than the level of the water in the container. As long as you hold the free end of the tube lower than the water in the container, water will continue to drain.

PEDAL POWER

ARE you using up too much energy when you bicycle? You are if your tires don't have enough air in them. Here's an experiment to show you how to increase your pedal power — without using any more energy.

You'll need:
a bicycle pump
some chalk
a hill

1. Make sure your tires are inflated to the proper pressure (it's often printed on the side of the tire itself).
2. Walk your bike to the top of a hill, give yourself a little push — just enough to start the bike rolling — and coast down the hill.
3. When your bike has rolled to a stop, mark the spot with the chalk.
4. Walk your bike back to the top of the hill and let air out of your tires, so they're only about half-inflated.
5. Start off with the same gentle push as before and coast down the hill. Do you think you'll go as far?

How does it work?
All objects (including your bike tires) resist sliding, moving, or rolling across other objects (the road); this resistance is called *friction*. Friction increases with the amount of contact between the objects. So the less air in a tire, the more it flattens out against the ground, and the more friction there is between the tire and the road. This makes it harder for the tire to roll, and the bike slows more quickly when you coast.

Increased friction also makes you work harder to pedal your bike, so you'll save a lot of your energy by making sure that your bike tires are sufficiently inflated. You can help save fuel energy, too, by checking the tire pressure on your family car. The correct pressure for safety and efficiency is printed on a sticker, usually inside the car door.

LOW-ENERGY BREAKFAST

HAT better time to start saving energy than at breakfast. Here's an egg-speriment that'll cut down on your family's energy use.

You'll need:
2 saucepans about the same size, and lids for both
a watch or clock
2 eggs

1. Pour equal amounts of water into each saucepan. Make sure both have enough water to cover an egg.
2. Cover one saucepan with a lid. Put the saucepans on separate burners of the same size on your stove, and turn them on to high.
3. Start timing to see which pot of water comes to a vigorous boil first. (You'll have to listen to the covered one to tell when it boils — lifting up the lid will spoil the experiment.) How much longer does it take for the second pot of water to boil?
4. When the two pots of water are boiling, place an egg in each one and cover both pots. Leave one burner on high heat and lower the other to simmer.
5. Time them for three minutes, then remove both eggs at the same time.
6. Break the eggs and see if one is more cooked than the other. Then eat the eggs.

How does it work?
Why does the covered pot boil first? Water molecules are always moving. But as water gets hot, the water molecules move around faster. They bump into others and make them move and bump into yet others, and this keeps happening until all the water molecules are rushing around and the water is boiling. As they move faster, the molecules near the surface escape from the water. Without a lid on the pot, these molecules disperse into the air, and their energy is lost from the water. But with a lid on the pot, those molecules are contained, and continue to stir up the rest of the molecules in the water, thus bringing it to a boil more quickly. Therefore you need less energy from the burners on the stove.

Why do both eggs cook at the same speed, even though one is cooked on higher heat? Water can't get any hotter than 100°C (212°F). At that temperature it becomes a gas, or water vapour. It doesn't matter whether the water is boiling ferociously or just gently bubbling — the temperature of the water is the same. The only difference is you're wasting a lot of energy to keep the water boiling hard. Enjoy your breakfast.

BODY TRICKS

ERE are two optical illusions guaranteed to fool you.

Illusion #1
Are those stairs on the floor or the ceiling? Are your eyes fooling you? No . . . your brain is.

How does it work?
Just as you had to learn to read — to make sense out of squiggles on paper — so you had to learn to see — to make sense out of rays of light hitting your eyes. Once your brain has learned the "rules" of seeing (for example, the farther things are from you, the smaller they look), it applies those rules to interpret everything you look at. But when an object or a drawing breaks the rules, or when it could be interpreted in different ways, your brain may give you wrong or confusing information, and that's what we call an optical illusion.

As you look at the staircase, you'll see it seem to flip upside down. The back wall becomes the front wall, and the stairs hang from the ceiling. That's because the drawing contains equal information for both interpretations, so your brain can't decide which way is correct. It keeps switching back and forth in the way it interprets the picture.

Illusion #2
Here's an illusion you can make for yourself.

You'll need:
a piece of stiff paper
a table

1. Fold the paper in half. Make the fold sharp by running your fingernail along it.
2. Set it on the table as shown, cover one eye, and stare at the folded edge. What happens?

How does it work?
By placing the paper in an open space and by covering one eye, you make it very difficult to judge depth correctly. Your brain becomes confused about which way the paper is folded, and tries different ways of seeing it.

TWO EYES ARE BETTER THAN ONE

ID you ever wonder why you need two eyes? One reason is that they help you see depth. To find out how different your depth perception would be with only one eye, try this experiment with a friend.

You'll need:
a cup
a penny

1. Put the cup on the table and stand about 3 m (9 feet) away.
2. Cover one eye. Have your friend hold the penny at arm's length above the cup, but slightly in front of it.
3. Watching only the cup and the penny, tell your friend where to move his arm so that the penny will fall into the cup when he drops it.
4. Tell him to drop the penny and see how close you came. Why are you such a bad shot?

How does it work?
Because our eyes are set apart from each other, they see everything from slightly different angles. So the images your brain gets from each eye are a little different from one another. By comparing the images, your brain can give you a three-dimensional picture which helps you judge distances. This is called stereoscopic vision. When you cover one eye, you no longer have stereoscopic vision and you see things in two dimensions, like a photograph. This makes judging distances much more difficult.

Fortunately, there are other clues to help you judge depth in real life, such as size, brightness, and position compared to other familiar objects. These are the clues that people use if they lose the sight in one eye. You can improve your one-eyed depth perception too. Try the penny drop test several times. You'll soon be hitting the cup quite easily.

61

TASTE TEST

 AN you taste the difference between a turnip and a carrot, or an apple and a potato? Think you can? Try this taste test with a friend.

You'll need:
small, peeled cubes of raw apple, potato, carrot, turnip, and even onion (all cubes should be about the same size)
a plate
a blindfold
a pen or pencil and paper

1. Spread the vegetable pieces on a plate.
2. Blindfold your friend so she can't tell what the vegetables are by sight and ask her to hold her nose.
3. Ask her to taste each vegetable cube one by one and guess what she's eaten. Write down what the cube really is and what she thinks it is.
4. Now it's your turn to take the taste test. At the end, compare your scores. Chances are both of you made a lot of wrong guesses.

How does it work?
Why was it so hard to tell one food from another of similar texture? The secret is inside your nose! The taste buds on your tongue — those tiny bumps on the surface — can only identify sweet, sour, salty, and bitter. The rest of the information about the taste of food comes from its characteristic odour. So without being able to smell, you can't taste!

Now you know why food has no taste when you have a cold in the nose.

Your Terrific Tongue
Your tongue has separate areas to taste sweet, sour, salty, and bitter.

The back of your tongue is where the taste buds for bitter are located. If you've ever tasted tonic water, for instance, you'll notice that the bitterness lingers at the back of your tongue.

Your sour taste buds are along the sides of your tongue, towards the front. Think of a lemon and you'll feel them getting ready to taste it.

Further back along the sides are your taste receptors for saltiness, and some saltiness tasters are on the tip of your tongue too.

Surprisingly, sweetness, one of the most popular tastes, can only be detected by the tip of the tongue.

HOT OR COLD TRICK

AN a bowl of water be hot and cold at the same time? Try this easy experiment and find out.

You'll need:
3 bowls

1. Pour cold water into one bowl, hot water into another, and lukewarm water into the third.
2. Dunk one hand into the cold water and the other into the hot water and leave them there for a minute or two.
3. Put both hands together into the lukewarm bowl. The lukewarm water will feel hot to one hand and cold to the other at the same time!

How does it work?
The hand that was in the cold water feels the medium temperature as hot; the hand that was in the hot water feels it as cold. You're experiencing sensory adaptation. That's what happens when any of your senses is exposed to the same strong sensation for a while. Your sense receptors get used to it and stop sending reports to your brain. That's why the good smell of dinner cooking is so powerful when you walk into the house, but fades away after you've been inside a few minutes. It's only when the sensation changes that you notice it again.

Sometimes your sense receptors can be fooled by dramatic changes and give you a false report, as in this experiment, where each hand feels the water as being the opposite to the temperature the hand was used to.

SEEING SOUND

ERE'S an unusual chance to see a sound.

You'll need:
a balloon
scissors
an orange juice or soup can with both
    ends removed
rubber bands
tape
glue
a tiny piece of mirror about 0.5 cm (½ inch) square
a flashlight

1. Cut the neck off the balloon and stretch the
   remaining part tightly over one end of the can.
   Hold the balloon in place with rubber bands and
   tape the edge of the balloon to the can to keep it
   from slipping.
2. Glue the piece of mirror (face out) to the stretched
   balloon, about a third of the way in from the edge of
   the can.
3. Now shine the flashlight onto the mirror at an
   angle, so that you can see a bright spot from the
   mirror reflected on the wall. If you don't have a
   plain wall to aim the spot at, use a piece of white
   cardboard as a screen.
4. Hold the can very still (or set it on a table, braced
   so it won't roll) and sing or shout into the open end
   of the tin. Watch the spot of light on the wall. Why
   does it vibrate quickly back and forth?

How does it work?
Sound is made by vibrations. When you sing or shout,
the air rushing from your lungs passes through your
vocal cords and makes them vibrate, producing
pressure waves that travel through the air, like ripples
in water. When these waves hit the stretched balloon,
they make it vibrate. This, in turn, starts both the
mirror and the light reflecting from it, vibrating.

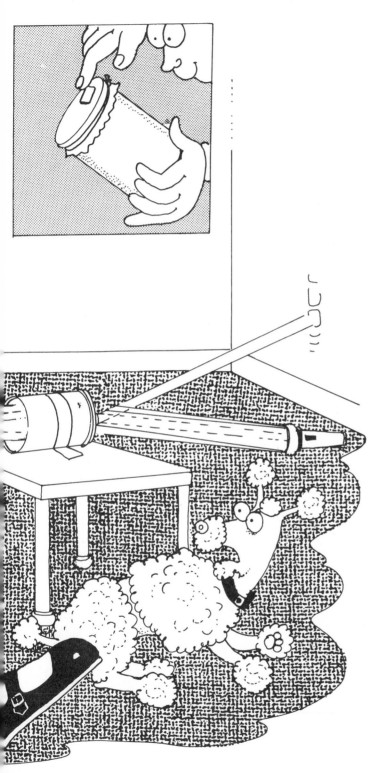

Your eardrum is a stretched membrane something like the balloon. When pressure waves strike the eardrum, it vibrates and your brain interprets those vibrations as sounds.

Human Voice vs Tape Voice
Remember how surprised you were when you heard your voice on a tape recorder for the first time? What makes it sound different? When you hear your voice on tape, you're hearing it after the sound waves have travelled through air. But usually you hear your voice through the bones of your head, and it sounds different. The way you sound on tape is actually closer to how others hear you.

Hearing Through Your Teeth
Find out what good "sound conductors" your head bones are.

You'll need:
a fork
a spoon

Hit the tines of the fork with the spoon and listen to the note produced by the vibrating fork. As soon as the sound fades, put the end of the fork handle between your teeth and bite firmly on it.

DID you know that you can make a hot and cold map on your body?

You'll need:
2 fine-tipped felt pens of different colours
a bowl
a nail

1. Draw a box about 2 cm square (1 inch square) on the back of your hand with one of the felt pens.
2. Fill the bowl with cold water and set the nail in it until it's cold to the touch.
3. Touch the tip of the cold nail to any spot inside the square. If it feels cold, mark the spot with one of the pens. Try other places inside the square, marking the ones that feel cold with the same colour. You'll probably have to keep putting the nail back into the water to keep it cold.
4. When you've tested the entire square for cold, heat the nail in hot water and, with the other colour,

mark where you feel heat. If you touch a marked cold spot with the hot nail, can you feel it? When you've finished marking off all the heat and cold receptors in that square, examine the map you've made. Were there more hot spots or cold spots?

How does it work?
Scientists have found that our bodies have separate spots, called receptors, for feeling temperatures that are hotter or colder than body temperature. The two different colours of dots on your hand give a map of your own hot and cold receptors. Other areas of your body may produce different maps. Try mapping small squares on your forehead, fingers, chin, palm, forearm, and the sole of your foot. You might find out the best way to pick up a snowball without feeling the cold!

THINGS TO MAKE

Why not whip up a batch of invisible ink, or churn some butter, or .... Read on for 13 other great things to make.

MAKING BUTTER

Do you like to spread coalesced fat droplets on your bread? If you do, you can make some sandwiches with the results of this experiment.

You'll need:
300 mL (½ pint) whipping cream
a small glass jar with a tight cover

1. Take the cream out of the refrigerator and let it stand for about 10 minutes so that it warms up slightly.
2. Pour the cream into the jar until the jar is one-third full.
3. Screw on the lid and make sure it doesn't leak.
4. Hold the jar in one hand and shake it in a figure-eight motion. (You'll have to keep shaking it for about 20 minutes, so it might be a good idea to have a helper available to take over in case your arm gets tired.)
5. Watch the cream change form. It will become foamy and almost look like it's whipping, but after a while, very tiny granules of butter will start to form.
6. When the granules are the size of apple seeds, stop shaking.
7. Carefully drain off the liquid — it's buttermilk and very good to drink.
8. Wash the butter granules in cold water to rinse off any remaining buttermilk.
9. Put the butter granules in a plastic or wooden dish and pack them together with a wooden spoon. If you like salty butter, add a little salt and squish it into the butter with your wooden spoon. Make sure it's worked in evenly.
10. Form your butter into any shape you like, then put it into the refrigerator to harden.

How does it work?
Cream is very tiny fat droplets permanently floating in water. If enough of those droplets can be forced together, they'll form globules of butter and separate from the water. This is called coalescing.

Mellow Yellow
You might notice that your butter is darker or lighter than the butter you usually get in the store. That's because the colour of pure butter depends on the kinds of cows the milk came from and what they ate. But when people buy butter, they expect it to be a uniform golden colour, so dairies usually dye it with food colouring.

MAKING SUGAR CRYSTALS

ID you ever wonder where the sugar goes when you stir it into tea? It doesn't disappear. To prove it, here's an experiment that'll let you get the sugar out of a hot drink in beautiful crystal form.

You'll need:
a small saucepan
250 mL (1 cup) water
375 mL (1½ cups) sugar, or more
a drinking glass
a long pencil
a piece of cotton string

1. Boil the water in the saucepan, turn off the heat, add the sugar, and stir. If all the sugar dissolves, add a little more and keep stirring until no more sugar will dissolve.
2. When the solution has cooled, pour it into the drinking glass.
3. Rub some sugar into the string so some crystals stick into it.
4. Tie one end of the string around the pencil and drop the other end into the solution. Rest the pencil on the rim of the glass.
5. Put the glass in a place where it will stay cool and undisturbed. (You mustn't touch it or lift it up!)
6. Leave it for a few days and watch what happens.
7. Eat the results.

How does it work?

To understand how you make the sugar reappear, you need to know what happens when the sugar dissolves. It doesn't really disappear, of course. It just breaks up into smaller and smaller pieces until you can no longer see it.

If you look at a sugar cube with a magnifying glass, you will see it is made up of lots of small crystals. These, in turn, are made up of tinier particles called molecules, the tiniest form of sugar that can exist. They are so tiny that you couldn't see them even with the most powerful microscope.

When you put your sugar in water, the sugar molecules break away from the crystal. This makes a sugar solution. The amount of sugar the water can hold depends on its temperature. Hot water can hold more than cold water.

As the solution cools, it becomes supersaturated. At this point, there is more sugar in the solution than can remain dissolved at the cooler temperature, and some of the sugar starts to come out of the solution and join the sugar crystals on the string. After a few days, your crystals should be quite large — big enough to eat! What you've done in this experiment is to reverse a process. You've taken a solution and caused the molecules to turn back into crystals.

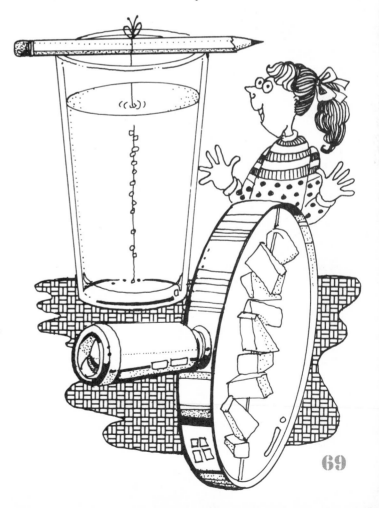

BUILDING A HOUSE

Ave you ever watched a house being built? Under the bricks or siding, there's a frame of wooden beams — the skeleton of the house. The frame gives the building its strength. It's not difficult to build a frame that would hold up a house. Test your construction skills by building with beams made of newspaper.

You'll need:
sheets of old newspaper (make sure everyone has read them!)
toothpicks
tape

1. Lie a sheet of newspaper flat on the floor. Place a toothpick across one corner and roll the newspaper tightly around the toothpick until the whole sheet is rolled. Fasten it with the tape. If you've rolled tightly enough, you'll have a long, strong, newspaper dowel that is very hard to bend. (If you want to shorten it, cut off the ends.)

2. Repeat the above process until you have a pile of newspaper beams.

3. When you're ready to start building, attach the beams together with tape. Start by outlining the shape of your building on the floor, using the beams. Then build up.

4. As you build, you might have to brace your frame with crossbeams.

5. Build your building as high as you can. Can you make it reach the ceiling?

SUPER STRUCTURES

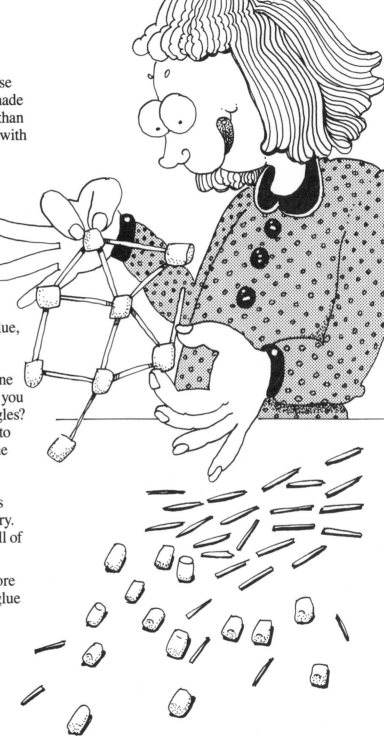

THE next time you see a house or a high-rise under construction, look at the shapes made by the frame. Some shapes are stronger than others. You don't need steel girders to experiment with shapes. You can raid the kitchen for building materials.

You'll need:
a package of toothpicks
miniature marshmallows (or modelling clay)
a hardcover book
2 chairs
5 quarters in a paper cup

1. Using the toothpicks as beams and the marshmallows (or balls of modelling clay) as glue, try to build a tall structure using nine marshmallows and 15 toothpicks.
2. Now try building with 15 marshmallows and nine toothpicks. Which structure is stronger? When you look at the stronger one, do you see more triangles?
3. Using 14 marshmallows and 20 toothpicks, try to make a structure that's strong enough to hold the book.
4. Here's a final challenge: Try building a bridge between two chairs 30 cm (1 foot) apart. Use as many marshmallows and toothpicks as necessary. When you're done, see if it will hold the cup full of quarters.

Note: If you want to make bigger, stronger and more permanent structures, attach the toothpicks with glue instead of marshmallows.

REINVENT THE CAMERA

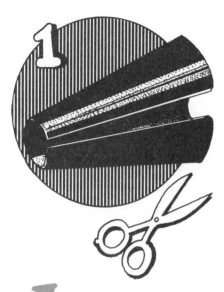

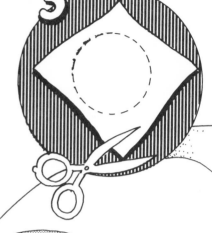

I F you darkened a room and drilled a small hole through one outside wall, the light coming through that hole on a sunny day would form a dim, upside-down image on the far wall. If you fastened a large enough sheet of photographic paper to the wall and left it there long enough, you'd get a photograph of the scene outside.

This may sound farfetched, but it's much the way the first photograph was made in 1826 by a French physicist named Nicephore Niepce, using a device called a camera obscura, which means "dark chamber."

Here's a way to make a small version of the camera obscura.

You'll need:
black construction paper
wax paper or tracing paper
an empty, frozen orange juice can
tape

1. Roll the construction paper into a cone and trim the wide end until it just fits inside the can opening.
2. Tape or glue the cone to keep it in shape.
3. From the wax paper or tracing paper, cut a circle the same size as the wide end of your paper cone.

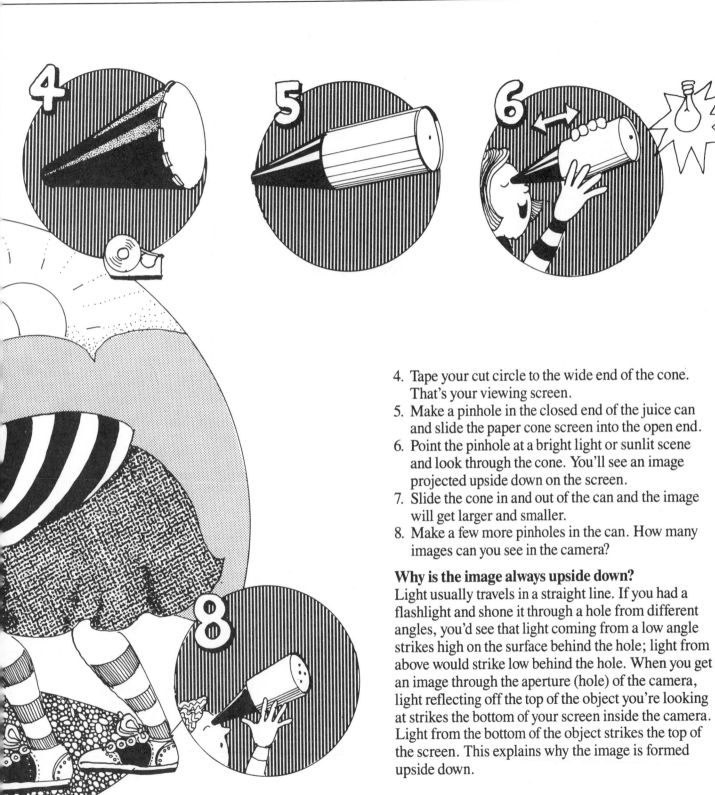

4. Tape your cut circle to the wide end of the cone. That's your viewing screen.
5. Make a pinhole in the closed end of the juice can and slide the paper cone screen into the open end.
6. Point the pinhole at a bright light or sunlit scene and look through the cone. You'll see an image projected upside down on the screen.
7. Slide the cone in and out of the can and the image will get larger and smaller.
8. Make a few more pinholes in the can. How many images can you see in the camera?

Why is the image always upside down?
Light usually travels in a straight line. If you had a flashlight and shone it through a hole from different angles, you'd see that light coming from a low angle strikes high on the surface behind the hole; light from above would strike low behind the hole. When you get an image through the aperture (hole) of the camera, light reflecting off the top of the object you're looking at strikes the bottom of your screen inside the camera. Light from the bottom of the object strikes the top of the screen. This explains why the image is formed upside down.

73

HERE'S a way you can see around corners or over the heads in a crowd. You just have to bend light. A job for Superman? No, you can do it yourself with this periscope.

You'll need:
scissors
a clean, empty milk carton
two pocket-sized mirrors
tape

1. Cut a hole in one side of the carton, near the top, and a similar hole in the opposite side, the same distance from the bottom.
2. Tape the two mirrors inside the box, facing one another as shown, making sure they're parallel to one another and slant across the box at a 45° angle.
3. When you have the mirrors secured inside, tape the top of the carton shut.
4. Take your periscope to a corner and hold it so just one hole is sticking out. Look through the other hole and you'll be able to see around the corner.

How does it work?
Light always bounces off a mirror at the same angle at which it hits. If it hits the mirror at 45°, it will reflect at 45°, enabling it to make the 90° turn around the corner. You can test this by shining a flashlight into the hole where you would look. If your mirrors are correctly angled, the light will shine out the other hole. Similarly, light reflecting off an object you're looking at will bounce off each mirror and into your spying eye.

INVISIBLE INK

ERE'S a way to write secret messages —
messages so secret, they're invisible.

You'll need:
vinegar or lemon juice for ink
a toothpick or paintbrush to write with
a piece of paper
a candle in a holder

1. Dip a toothpick or paintbrush into your invisible
   ink and write your message on a piece of paper.
   When the message is dry, the paper will look blank.
2. To read the secret message, pass the paper back and
   forth over the flame of the candle. Ask your parents
   to help with this and be sure you don't let the paper
   catch fire. Gradually the writing will appear.

How does it work?
That heat from the flame causes a chemical change in
the dried ink. The portion of the paper which
absorbed the vinegar or lemon juice chars at a lower
temperature than the untreated paper, so the writing
shows up as a faint brown scorching.

Secret Writing Tips
1. If you're using a toothpick for a pen, write with
   the round end. That way you won't scratch or
   tear the paper.
2. Don't press too hard when you're writing your
   message. If you do, you'll make an impression
   on the paper that can be read even without
   decoding it over a flame.

AMAZING HOMEMADE AIRPLANES

IRED of the same old paper airplane designs? Try these two unusual flying wonders.

Straw Plane

You'll need:

one strip of paper 1.5 cm × 9 cm (½ inch × 3½ inches) long

one strip of paper 2 cm × 12 cm (¾ inch × 4¾ inches) long

a regular-sized plastic straw

cellophane tape

1. Make a loop out of each strip of paper, overlapping the ends and taping them inside and outside the loop. The overlapped ends will form a pocket into which you can slip the straw.
2. Put one loop on each end of the straw by slipping the straw through the pockets you've made.
3. Experiment with the loops in different positions along the straw. Try it with the loops on the top and the bottom and take turns putting each loop at the front.

How does it work?

Paper airplanes — even the odd looking one you've just made — fly using the same principles as real airplanes. When they're moving, the shape and angle of their wings cause the air to move faster over the wing than under it. This reduces the pressure of the air above the wing, increases the pressure underneath the wing, and the plane is held up by the difference.

A real airplane must race down the runway to get the air moving fast enough past the wings to create enough difference in air pressure to lift it, and then must stay above a minimum speed while in the air. A helicopter, on the other hand, moves just its wings — the whirling rotors. This forces the air past them at a speed that's enough to lift it off the ground, or slow its descent.

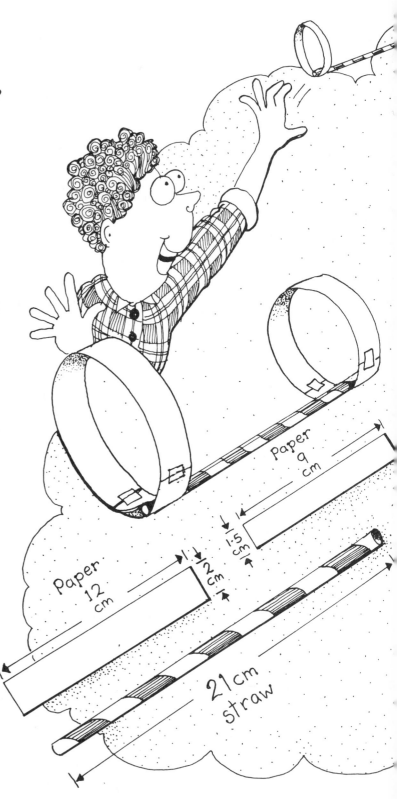

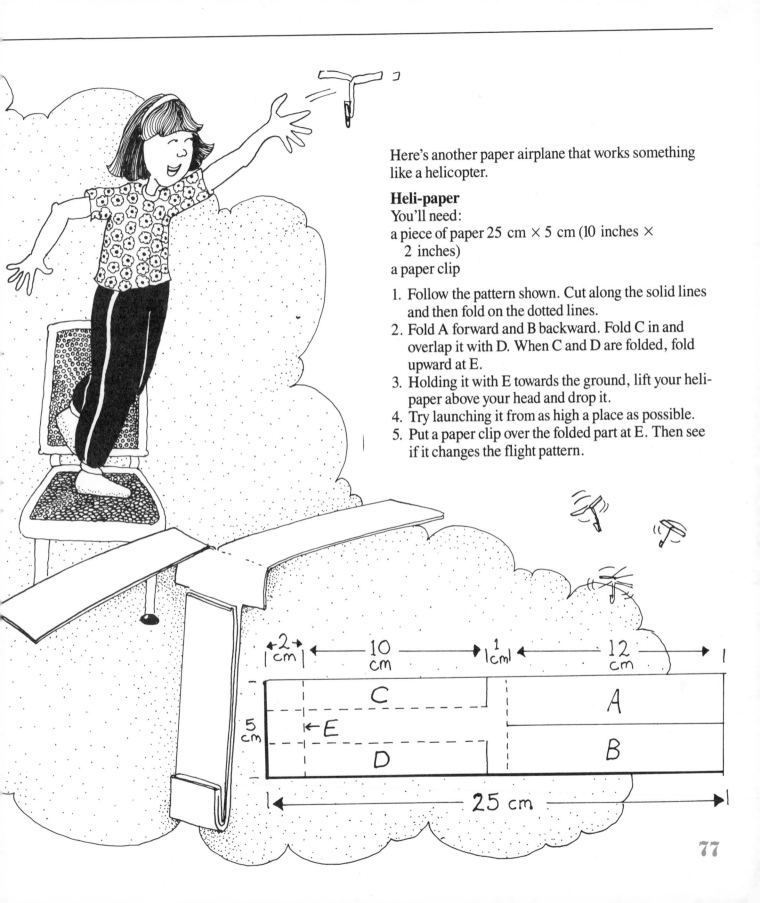

Here's another paper airplane that works something like a helicopter.

Heli-paper

You'll need:
a piece of paper 25 cm × 5 cm (10 inches × 2 inches)
a paper clip

1. Follow the pattern shown. Cut along the solid lines and then fold on the dotted lines.
2. Fold A forward and B backward. Fold C in and overlap it with D. When C and D are folded, fold upward at E.
3. Holding it with E towards the ground, lift your heli-paper above your head and drop it.
4. Try launching it from as high a place as possible.
5. Put a paper clip over the folded part at E. Then see if it changes the flight pattern.

MAGNIFYING EXPERIENCE

CLEAR PLASTIC WRAP
STRING OR TAPE
WATER
HOLE
OLD PAIL

SOME things are too small to see no matter how closely you look at them. That's why people invented magnifying glasses and microscopes — to make things appear bigger. Make your own version of a microscope and see what's hidden from view.

You'll need:
scissors
an old plastic pail
clear plastic wrap
water
string, tape, or a thick rubber band

1. Cut two or three fist-sized holes in the side of the pail, near the bottom.
2. Stretch the plastic wrap loosely across the top of the pail and fasten it securely around the side with the string, tape, or rubber band (or use all three together).
3. Pour water into the hollow formed by the plastic wrap until it is filled.
4. In a brightly lit room, put the object you're studying into the pail, through one of the holes.
5. Look down through the water into the pail. Move the object closer towards you and farther away until you've found the point where the object appears largest. Now find the point where the object is in

the sharpest focus. Are they the same point?
6. Try to find the designer's initials on the back of Canadian pennies, quarters, and nickels. (Here's a clue for finding it on the quarter: look near the edge just in front of the caribou's chest.)

How does it work?
A lens bends light both as the light enters and again as it leaves. The material the lens is made of determines the angle at which the light bends. In this case, you've made a water lens. Reflected light spreading out from the object you're looking at hits the lens and is bent back to your eye (see the diagram). Your eye sees the light as though it came on a straight line from the object (the dotted lines in the diagram), and it appears as though you're looking at a much larger object a comfortable distance away.

Other ways of seeing
When scientists want to see something as tiny as germs, they use light microscopes with specially shaped and highly polished lenses. They also use electron microscopes which send beams of electrons through objects to produce pictures of things almost as small as atoms.

DRAFT DETECTOR

YOU can be a draft detective. Most buildings leak air through tiny holes and cracks that are hard to find. Track down those cold rivers of air that chill your back when you lie on the floor. You can do it! All you need is a draftometer.

You'll need:
scissors
a strip of tissue paper or tissue
tape
a long pencil

1. Tape one end of the paper or tissue along the length of the pencil.
2. Blow gently on the tissue to make sure it's well taped. See how easily it responds to air movement!
3. Hold the draftometer near the edges of the windows and doors in your house. (If your home is heated by a forced air furnace, wait until the fan is off before you use your draftometer.)
4. If your home has a fireplace, hold the draftometer in front of it and test what happens with the damper open and closed.

   Once you've found all the air leaks, it's a simple job for the family handyperson to plug them up.

What makes a draft?
Air leaks through the cracks in your house. In the winter, this lets cold air in from the outside and allows the hot air to escape. Therefore, the inside temperature cools down and energy is wasted. In the summer, the same leaks let the cool, house air out and allow summer's heat to pour in.

Cold air is denser than warm air; therefore it tends to fall — that's why you usually feel cold drafts along the floor. Warm air moves in to take the place of falling cold air. So when you open your refrigerator door, you not only lose the nice cold air that is inside, but warm air flows in and the refrigerator has to work even harder to cool it off.

Check this out
Open the refrigerator door a crack and hold your draftometer at the bottom, in front of the opening. Close the door and try it again, this time holding your draftometer at the top. Which way does it blow?

MAGNETIC IMAGES

YOU'VE probably drawn with crayons, paints, and coloured pencils, but have you ever painted patterns with pieces of metal? Here's a fun and easy way to create a picture to hang on your wall.

You'll need:
two magnets
fine steel wool
stiff paper or light cardboard (file cards are ideal)
old scissors
hair spray or clear, plastic fixative spray
some metal objects such as paper clips, nails, etc.

1. Cut the steel wool into tiny pieces with the scissors.
2. Put the magnet on a table and place the sheet of paper over it.
3. Generously sprinkle some steel wool filings on the paper. You'll see that the filings have collected into a pattern.
4. Tap the paper gently to change the pattern or move the paper over the magnet until you're satisfied with the pattern.
5. If you have two magnets, put them both under a piece of paper and watch what happens to the filing patterns when you move the magnets around. Try putting the magnets end to end, top to end, and top to top to see what you can create.
6. If you have any other metal objects handy, put them under the paper to see what patterns they cause in the filings. Touch these objects to the magnets to see if you can magnetize them to help make your picture even more dramatic.
7. When you finish your metal paintings, clean up any scattered or fallen filings by covering your magnet with a piece of paper, then passing it closely over your work area.
8. When you have a pattern of filings that you want to keep, spray it several times with hair spray, or clear, plastic fixative, while the magnet is still in place. Let it dry between sprayings.

How does it work?

When you sprinkled the filings onto the paper on top of the magnet, you probably noticed that they formed a pattern. And when you tapped the paper gently, the filings collected even more definitely along the pattern lines.

These lines indicate the field of force that reaches out from the magnet. When the paper is moved over the magnet, the filings shift to follow the force field.

If you've played with magnets before, you may have already discovered that they have two poles, usually called north and south. If you put the north pole of one magnet near the south pole of another, they pull towards, or attract, each other. But if you put the two north poles or the two south poles together, they push one another away. You can see the difference in your magnet drawings.

When you touched the other metal objects to the magnet, you may have discovered that some of them became magnets themselves, as long as they remained in contact with your original magnets. Some even stayed magnetized after they were separated from the original magnet. You can test this by putting the separate metal objects under a paper covered with metal filings.

INSTANT ARTIST

Do you have a favourite picture or photograph that you'd like to copy? Perhaps you'd like to make it larger or smaller than the original. Here's a bright way to do it.

You'll need:

a picture to copy (a clear, simple one is best to start with)

a 20 cm × 25 cm (8 inch × 10 inch) piece of glass (the glass from an old picture frame is good)

a piece of white paper

a pencil

a desk lamp

1. Put the picture and the piece of white paper side by side on a table.
2. Place the lamp beside the picture and shine the light directly on it. Hold the glass upright between the blank paper and the picture.
3. From the picture side, look through the glass and you'll see an image of the picture on the blank sheet. You may have to move your head until you find the best viewing spot for the image.
4. Hold the glass and your head steady and you can trace the image on the paper.

How does it work?

How can you see the picture on the paper when it isn't really there? Light shines on the picture and is reflected from it. Most of it just travels away in straight lines in all directions, but some of it bounces off the glass and into your eye.

When this reflected light enters your eyes, you see the picture as though it were on the paper. That's why moving your head can make the picture appear larger, or smaller, or make it disappear altogether.

What you've made is a simple version of the camera lucida, a device invented in 1807. It has often been used by both artists and scientists to enlarge or reduce drawings.

FABULOUS FLYWHEEL

AVE you ever used a potter's wheel or a treadle sewing machine? When you pump the pedal, you turn the flywheel which is the "energy control centre" of the machine. You can make your own flywheel very simply — and then experiment to see just how much energy control you can have.

You'll need:
one large button
1 m (1 yard) of strong thread or fine string

1. Pull the thread through the button's hole (use diagonal holes in a four-hole button). Make sure the thread isn't tangled. Tie the ends together.
2. Slide the button to the middle of the looped string. Take one end of the loop in each hand and twirl the button around about 20 times in one direction, until the thread is all twisted.
3. Pull the ends apart and watch what happens to the button. Hold your hands still after you've pulled once and keep watching the button.
4. Alternate pulling and relaxing to keep the button spinning back and forth for a long time.
5. While you're spinning the button, try pulling the string harder sometimes, more gently at other times. Watch what happens to the button's speed when you change the strength of the pull.

How does it work?
When you wind the string, you are storing mechanical energy in it. When you pull the string, it straightens out and the stored energy is transferred to the button, making it spin. The bigger the button, the more energy it can store, and the longer it will spin. As it spins, it transfers the stored energy back to the thread. If you just kept your hands still, eventually the button would stop spinning, because a little energy is lost every time it is transferred from thread to button and back again. But when you pull on the thread, you add more of your own energy, so the button will keep spinning as long as you keep pulling.

Flywheels are used in many situations to provide a constant output of energy even when the supply of energy varies. When you pump the treadle on a potter's wheel or a treadle sewing machine, you turn a flywheel which, in turn, spins the clay platform or runs the sewing needle. If you let up for a minute on the treadle, or if you pump really hard, the flywheel absorbs the changes in energy supply, but keeps on spinning at a constant rate, turning the machine at a steady speed.

BUBBLE, BUBBLE . . . WITHOUT TROUBLE

ERE are three unusual bubble experiments. Before you get started on them, mix up a batch of superbubbles.

Superbubbles
Mix together and gently shake the following:
6 glasses of water
2 glasses *clear* dishwashing liquid detergent
(Joy® is the best)
1 - 4 glasses glycerine — 4 is best, but because of the cost of glycerine, you may wish to use less than 4 glasses (You can buy this at any drugstore.)

Monster Bubbles
These bubbles are so big they look as if a monster made them.

You'll need:
a thin wire coat hanger
superbubble mixture

1. Undo the coat hanger and twist it into a big circle.
2. Dip it into the superbubble goop and blow. With a little practice, you'll soon be blowing huge bubbles.

How does it work?
When you add soap to the water, you loosen the hold that the water molecules have on each other. This makes the water more "stretchable" so you can blow bigger bubbles. The glycerine makes the bubbles last longer than usual.

Slow Breakers
Most bubbles pop instantly when you puncture them, but not these.

You'll need:
superbubble mixture
a funnel
a piece of string about 20 cm (9 inches) long
a sharp pencil

1. Tie a piece of string around the stem of the funnel so that the string hangs down below the funnel.
2. Tie a loop in the loose end of the string.
3. Dip the big end of the funnel, string and all, into the superbubble goop.
4. Remove it and blow a big bubble by blowing into the stem of the funnel. The string will lie along the outer surface of the bubble.

5. Using a sharp pencil, poke a hole in the bubble through the loop of the string and watch what happens. First, the loop of string will stretch into a perfect circle. Then the bubble will slowly collapse.

How does it work?
Why doesn't the bubble pop instantly when you puncture it? The string prevents a long tear from developing in the soap film. The bubble can collapse only as quickly as air can escape through the small hole inside the string.

Soap Film
Soap film is a bubble with no air in it. As you'll see by trying a couple of experiments, soap film is pretty interesting too.

You'll need:
superbubble mixture
a thin wire coat hanger

1. Undo the coat hanger and bend it into a rectangular shape.
2. Dip this frame into the superbubble goop (you

might have to pour the bubble mixture into a flat pan so that you can dip the whole frame in at once). Hold the framed soap film up to the light. You'll see a rainbow of colours. This is caused by light reflecting from the front and back surfaces of the film.
3. Now, dry the frame and tie string to it in the pattern shown. Don't stretch it tight; let it hang limply.
4. Dip the frame, string and all, into the superbubble goop. Lift it out and prick a hole in the centre loop. What happens to the shape of the centre loop?

How does it work?
Why does the string form a perfect circle? Soap film always shrinks to its smallest possible area. For the soap film to be as small as possible in area, the loop of string has to be as large as possible in area. A circle is the shape that lets the loop of string cover the most area.

Now that you've solved that mystery, try making a hole in a different part of the frame. Can you see the outline of part of a circle? Try bursting the sections one part at a time and see what happens to the string.

SCIENTIFIC CENTREPIECE

HERE'S a scientific centrepiece that's so unusual it's bound to be a conversation piece too.

You'll need:
a tall glass jar or pitcher
some white vinegar
baking soda
vegetable dye
mothballs (available at most hardware stores)

1. Fill the pitcher with water, then very slowly stir in about three teaspoons of vinegar and two teaspoons of baking soda. The liquid will start to fizz. (You might have to adjust the amount of vinegar and baking soda to the size of your pitcher. Just be sure there's always three parts vinegar to every two of baking soda.)
2. Add a couple of drops of vegetable dye — not too much — and drop in a few mothballs. The mothballs will sink to the bottom at first, but then they'll start to rise. When they reach the top of the liquid, they sink again and they'll keep doing this at least long enough for you to have dinner.

How does it work?
Your unusual centrepiece is caused by a chemical reaction. The vinegar is an acid and the baking soda is a base. When you combine them, you cause a chemical reaction that produces carbon dioxide bubbles.

Bubbles tend to gather on surfaces, as you may have noticed when you've put a straw into a soft drink. They gather in great numbers on the mothballs. Since a mothball isn't very heavy, the bubbles clinging to it soon lift it to the surface. When it reaches the top, the bubbles break, the mothball sinks back to the bottom, and starts collecting bubbles again.

The vegetable dye: that's just to make it pretty! You could make several containers of mothballs and dye them different colours for a larger centrepiece.

By the way, you can get the same effect by dropping salted peanuts into a soft drink!

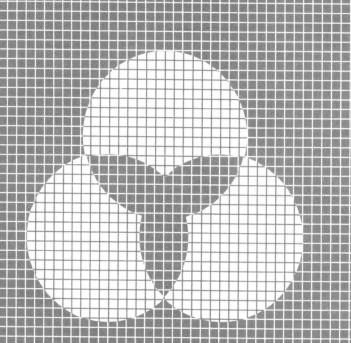